AS/A-LEVEL
STUDENT GUIDE

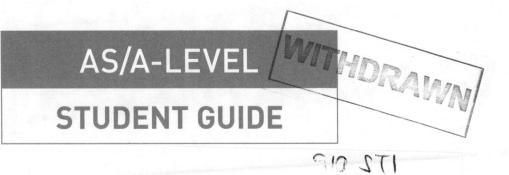

OCR

Geography
Climate change/Disease dilemmas/Exploring oceans/ Future of food/Hazardous Earth

Peter Stiff, David Barker and Helen Harris

D0416314

MPANY

Hodder Education, an Hachette UK company, Blenheim Court, George Street, Banbury,
Oxfordshire OX16 5BH

Orders

Bookpoint Ltd, 130 Milton Park, Abingdon, Oxfordshire OX14 4SB

tel: 01235 827827

fax: 01235 400401

e-mail: education@bookpoint.co.uk

Lines are open 9.00 a.m.–5.00 p.m., Monday to Saturday, with a 24-hour message answering
service. You can also order through the Hodder Education website: www.hoddereducation.co.uk

© Peter Stiff, David Barker and Helen Harris 2017

ISBN 978-1-4718-6414-8

First printed 2017

Impression number 5 4 3

Year 2021 2020 2019 2018

Cover photo: Richard Carey/Fotolia

Typeset by Integra Software Services Pvt Ltd, Pondicherry, India

Printed in India

Hachette UK's policy is to use papers that are natural, renewable and recyclable products and
made from wood grown in sustainable forests. The logging and manufacturing processes are
expected to conform to the environmental regulations of the country of origin.

Contents

◼ Getting the most from this book

Exam-style questions

Sample student answers

Practise the questions, then look at the student answers that follow.

Commentary on sample student answers

Read the comments (preceded by the icon **e**) showing how many marks each answer would be awarded in the exam and exactly where marks are gained or lost.

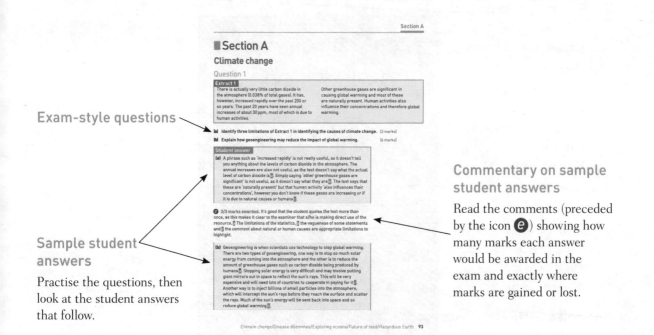

■ About this book

Much of the knowledge and understanding needed for AS and A-level geography builds on what you have learned for GCSE geography, but with an added focus on geographical skills and techniques, and concepts. This guide offers advice for the effective revision of the five AS and A-level option topics: Climate change, Disease dilemmas, Exploring oceans, Future of food and Hazardous Earth.

The external exam papers test your knowledge and application of these aspects of physical and human geography. More information on this is given in the Questions & Answers section at the back of this book. To be successful in these topics you have to understand:

- the key ideas of the content
- the nature of the assessment material — by reviewing and practising sample structured questions
- how to achieve a high level of performance within them

This guide has two sections:

Content Guidance: this summarises some of the key information that you need to know to be able to answer the examination questions with a high degree of accuracy and depth. In particular, the meaning of key terms is made clear and some attention is paid to providing details of case study material to help to meet the spatial context requirement within the specification. Students will also benefit from noting the **Exam tips** that provide further help in determining how to learn key aspects of the course. **Knowledge check questions** are designed to help learners to check their depth of knowledge — why not get someone else to ask you these!

Questions & Answers: this includes some sample questions similar in style to those you might expect in the exam. There are sample student responses to these questions as well as detailed analysis, which will give further guidance in relation to what exam markers are looking for to award top marks.

The best way to use this book is to read through the relevant topic area first before practising the questions. Only refer to the answers and examiner comments after you have attempted the questions.

Content Guidance

■ Climate change

How and why has climate changed in the geological past?

The Earth's climate has always fluctuated but usually over very long timescales and without the influence of human activities.

The Earth's climate is dynamic

Methods used to reconstruct past climates

A range of evidence is available to inform us about what the climate was like in the past.

- **Marine sediments:** fossil shells of foraminifera reveal the temperatures in which they formed.
- **Ice cores:** cylinders of ice removed from ice sheets contain bubbles of gas and we can measure the relative proportions of different gases, which mark climate changes.
- **Lake sediments:** cores from lake beds contain pollen grains, spores and diatoms (algae), which reveal what plant communities existed in the past.
- **Tree rings:** the width of each ring reveals how quickly the tree grew, which reflects past temperatures.
- **Fossils:** plants and animals are sensitive to particular climate conditions. Fossils indicate the climate in which they lived.

> **Foraminifera** Ancient single-celled organisms with shells, found in all marine environments.

Past climates reveal periods of greenhouse and icehouse Earth

The Earth's climate has altered from that of a greenhouse, when carbon dioxide levels, temperatures and sea levels were high, to an icehouse, when the opposite conditions existed.

Long-term changes

About 100 million years ago, during the mid-Cretaceous period, much of the world was subtropical, polar ice caps didn't exist and carbon dioxide levels were some five times higher than today. Then major tectonic changes occurred, which moved continents and affected ocean circulation. The Earth's energy budget altered, resulting in a cooling down.

> **Exam tip**
>
> Make sure you can discuss long-term changes to climate in terms of what happened, when and why.

Antarctica's glaciation

Fossils from Antarctica reveal that subtropical conditions existed 40 million years ago. Around 35 million years ago, rapid cooling of the Earth occurred, which culminated in ice sheets accumulating on Antarctica. Three factors were responsible:

1 Antarctica moved towards the South Pole, away from Australia and South America. This allowed an ocean current to flow right around the continent and prevent warmer water further north from coming south.

2 The South Sandwich Island's submerged volcanic arc affected ocean currents, preventing warmer flows from further north reaching Antarctica.

3 Carbon dioxide levels fell from 1,000–1,200 parts per million (ppm) to 600–700 ppm.

Quaternary glaciation

During the past 2.6 million years, the climate has altered between conditions as warm or even slightly warmer than today (**inter-glacials**) to long cold episodes (**glacials**).

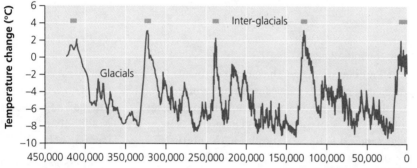

Figure 1 Glacials and inter-glacials in the late Quaternary period

Inter-glacial A period of warmer average global temperatures lasting approximately 10,000–15,000 years.

Glacial A period of about 100,000 years marked by colder temperatures such that continental glaciation occurs in the middle and high latitudes.

Over the past 450,000 years, four glacials and four inter-glacials occurred (see Figure 1). The most recent glacial, the Devensian, was at its maximum some 20,000 years ago. Ice sheets 1–1.5 km thick extended across the British Isles, leaving just the southern region ice-free but in conditions similar to northern Siberia today.

The Holocene

The Holocene (an inter-glacial) began approximately 11,700 years ago. Its climate included both cool and warm periods. About 6,000 years ago, average global temperature was 1°C to 2°C higher than today. The period 1100–1300 was relatively warm, while the years 1550–1850 are known as 'the Little Ice Age' as Europe experienced some very cold winters. Overall, ice has retreated and sea levels risen by some 100–120 m since the Devensian.

How natural forcing has driven climate change

Natural forcings are factors that result in the climate changing. They can either come from outside the Earth or from the Earth itself.

External forcings

Serbian astronomer Milankovitch theorised long-term cycles that affect the amount of solar radiation reaching the Earth's surface, and the spatial and temporal distribution of that energy. There are three types of Milankovitch cycle:

1 **Tilt of Earth's axis:** this varies between 22° and 24.5° over a period of about 40,000 years. The closer to 22°, the fewer seasonal differences there are (summers are cooler, winters are milder). Snow and ice do not melt during the summer and, even with milder winters, can expand. Positive feedback occurs, increasing the amount of radiation reflected, which further lowers temperatures. This forcing correlates with the pattern of glacials and inter-glacials.

2 **Earth's movement on its axis:** as the Earth spins on its axis, it wobbles slightly. This is due to gravitational influence from the Moon and Jupiter, occurring around every 22,000 years. The effect alters the amount of solar radiation the Earth receives. Northern hemisphere winters are milder when the Earth is particularly close to the sun, but summers are cooler. Similar climate change to that experienced during the tilting of the Earth's axis occurs.

3 **Earth's orbit around the sun:** this varies from nearly circular to very elliptical. A very elliptical orbit means that there is a significant difference (about 30%) in solar radiation received between winter and summer. Glacial periods correlate to very elliptical orbits.

Internal forcings

Examples of internal forcings include the following:

■ **Plate tectonics:** starting some 250 million years ago, the super-continent of Pangaea began to break up. As continents drifted, ocean currents shifted, altering the distribution of solar radiation absorbed by the oceans. The changing distribution of land altered the balance between incoming and outgoing solar radiation. For example, as more land moved to higher latitudes in the northern hemisphere, permanent land-based ice cover increased. Positive feedback occurred as global albedo increased, forcing global cooling.

■ **Volcanic gases:** volcanic eruptions can eject vast quantities of ash and gas into the atmosphere. Ash tends to dissipate in weeks to several months, but reflects solar radiation directly out to space. Ejected sulphur dioxide (SO_2) forms sulphate aerosols, which also reflect solar radiation into space. Some volcanic gases act as greenhouse gases. The more material erupted, the greater the impact on climate, but the effects only last for several years.

■ **Natural greenhouse gases (GHGs):** naturally occurring GHGs include water vapour, carbon dioxide (CO_2), methane (CH_4) and nitrous oxide (N_2O). More CO_2 and CH_4 lead to warmer conditions, and vice versa. GHGs such as CO_2 vary naturally due to several factors. For example, due to the uplift of mountain chains such as the Andes and Himalayas, rainfall increased. Chemical weathering via carbonic acid in rainwater also increased. Vast quantities of CO_2 were then transferred from the atmosphere to the long-term sink of marine carbonate sediments. The increase in nutrients from eroded and weathered rock in the oceans led to an explosion of the phytoplankton population, which removed CO_2 from the atmosphere. The carbon was buried in sea-floor sediments when these organisms died and sank.

■ **Solar output:** solar radiation varies and can be assessed using sunspot evidence. The reliable record of sunspot activity from the past 400 years shows a positive correlation between sunspot numbers and levels of solar energy emitted. A regular 11-year cycle in sunspot activity exists. However, the difference between maximum and minimum parts of the sunspot cycle gives roughly 0.1% variation in solar

Albedo The proportion of sunlight reflected from a surface.

Sulphate aerosols Minute particles that form from sulphur dioxide gas, which then convert to small droplets of sulphuric acid. The particles stay in the atmosphere for around 2 years.

Phytoplankton Tiny photosynthesising marine organisms living in surface waters.

Sunspots Large areas of intense magnetic activity on the sun's surface, which result in highly energetic solar flares shooting into space.

output, too little to alter global climate significantly. Over centuries, sunspot activity is more variable. The 'Maunder minimum' period (approximately 1650 to 1720) correlates less sunspot activity with severe winters in Europe.

It is important to note that climate models combining all natural forcing factors do not match the observed changes in temperatures, in particular the global warming of the past 50 years.

How and why has industrialisation affected global climate?

Global human activity is now so dominant that environmental change is seen as largely human-driven. This has led to the view that the Holocene has ended and we are now experiencing a new geological age: the **Anthropocene**.

The influence of humans on the climate system

Evidence that the world has warmed since the late nineteenth century

Increases in temperatures

Sea surface temperatures (SSTs), atmospheric and land surface temperatures have all risen. Global surface temperature increased by 0.74°C over the course of the twentieth century. All readings show sharp increases in temperature since 2000 — the ten hottest years on record have all occurred since 1998, nine of these since 2000.

Shrinking ice sheets and valley glaciers

In Greenland melting outpaces accumulation. The situation in Antarctica is more complex due to its three separate ice sheets. The Antarctic Peninsula ice sheet is melting and a large part of the West Antarctic ice sheet has lost mass, but no clear trend has emerged for the East Antarctic ice sheet.

In mountainous regions (the Alps, the Himalayas) most valley glaciers have been melting. More meltwater results in the acceleration of glacier flow, bringing more ice more quickly to lower altitudes, where temperatures are warmer. For glaciers flowing into the sea, faster flows lead to more ice breaking off at the coast and floating away as icebergs, which then melt.

Rising sea levels

Sea levels have been on the rise since the mid-nineteenth century. For most of the twentieth century, the average rise was 1.0–2.5 mm/year but satellite readings suggest this has increased to 3.0 mm/year. Two factors are responsible.

1 **Thermal expansion of water:** as SST increases, water density decreases and water expands in volume.

Knowledge check 1

What is the difference between external and internal natural forcing?

Anthropocene The term given to the current geological age. It is recognised as being when human activities started to have significant global impact on the Earth's geology and environment.

Exam tip

Websites such as those maintained by NASA, NOAA and the National Snow and Ice Data Center offer detailed and updated material on ice melt. Use these to keep up to date with the latest figures and give your answers authority.

2 **Transfer of land ice to the sea:** as temperatures rise, ice melts and water flows back to the sea.

To date, about half the sea-level rise is due to thermal expansion of water. Greenland and Antarctic ice melt accounts for some 14%. The rate and extent of ice melt is causing concern, as it is one of the biggest unknowns of global warming.

Increasing atmospheric water vapour

Water vapour is a natural GHG and accounts for some 50% of solar energy absorbed. The amount of atmospheric water vapour is directly related to temperature and rates of evaporation. A warmer world results in greater evaporation and more moisture in the atmosphere. Water vapour, however, does not last long in the atmosphere, since parts of the water cycle operate relatively quickly.

Models suggest that water vapour amounts may double by 2100. Due to positive feedback, roughly twice as much warming will occur than if the water vapour level remained constant.

Decreasing snow cover and sea ice

Snow measurements taken in early spring allow for the maximum amount of snow accumulation over winter. Since the mid-1960s, the area of snow in the northern hemisphere has declined at 2% per year. Sea ice in the Arctic Circle has reduced by about 0.8% per year in summer and 0.3–0.4% in winter. Its thickness has reduced from nearly 4 m in the early 1980s to 1.9 m in the early 2000s.

Sea ice around Antarctica has a more complex set of changes. At the end of the southern hemisphere winter in 2016, sea ice extent was the second lowest on record. However, the trend during the late twentieth and early twenty-first centuries has been of slight increases of about 0.2% per year. Combining Arctic and Antarctic data shows that there is a steady decline in sea ice overall.

How have anthropogenic GHG emissions changed since pre-industrial times?

Anthropogenic GHG emissions are chiefly those that originate in human activity.

Carbon dioxide (CO_2)

Up to about 1800, atmospheric CO_2 levels were comparatively stable at around 280 ppm, but by the late 1950s had risen to approximately 316 ppm. Since then CO_2 concentrations have continued to rise, reaching 390 ppm in 1990 and passing the 400 ppm level in 2015. The rate at which CO_2 levels are rising has been increasing. Half the increase since 1800 has occurred since 1960.

Methane (CH_4)

Methane is produced when organic matter decays in anaerobic (oxygen-free) conditions. Until around 1800, methane levels were about 700 parts per billion (ppb). During the second half of the twentieth century, levels rose from around 1,000 ppb to 1,500 ppb, and are now close to 1,750 ppb.

Knowledge check 2

What factors are responsible for sea-level rise?

Exam tip

Practise drawing a systems diagram to illustrate the positive feedback operating in the relationship between increasing temperatures and the amount of water vapour. You could use this in an essay response.

Exam tip

It is important to stress in any discussion about carbon dioxide that the rate of change in levels of carbon dioxide in the past 100 years is comparable to the natural variations that took thousands of years to occur.

Methane Occurs naturally, e.g. in wetlands. Human activity also generates methane, e.g. agriculture (livestock, rice paddy fields), landfill, coal and oil extraction.

Nitrous oxide (N_2O)

Nitrous oxide has natural and anthropogenic sources, such as burning fossil fuels and nitrogen-based fertilisers. Its atmospheric concentration rose steadily during the industrial period and is currently at roughly 50 ppb, which represents a 16% increase during that period. This level is significant as N_2O persists in the atmosphere for up to 150 years and is some 200–300 times more effective in trapping heat than CO_2.

Halocarbons: CFCs, HCFCs and HFCs

These gases rarely occur naturally so are almost entirely the result of anthropogenic activities. Their concentration levels are very small when compared with carbon dioxide, but they are some 3,000–13,000 times more effective at trapping heat than is carbon dioxide. Governments have largely controlled CFC use and there have been moves to reduce other halocarbons. However, their persistence in the atmosphere (many decades and up to 400 years) means they will continue to act as GHGs.

Why have anthropogenic GHG emissions risen?

Population growth

In 1800, the global population was 1 billion. In 2016, it passed 7.4 billion and is expected to reach just over 11 billion by 2100. Not only does this lead to an increase in carbon footprints, the vast majority of people are increasing their carbon output through rising standards of living. For example, the production of food, clothes, clean water and housing all release GHGs into the atmosphere.

Land-use changes

Large areas of forest have been cut down, releasing carbon but also reducing the Earth's capacity to absorb carbon. Draining wetlands, increasing the area under cultivation, higher livestock numbers and increased use of landfill account for the release of approximately one-third of all GHGs.

Energy demand

The shift in energy supply to fossil fuels powered the industrial period. We still rely on fossil fuels to supply some 87% of total energy, for example in energy supply, manufacturing, transport, agriculture and domestic demand (see Figure 2).

How has the balance of anthropogenic GHG emissions changed?

From 1850 to 1960, Advanced Countries (ACs) were responsible for most GHG emissions. The USA was the largest source (some 28% since 1850), but the last 40 years has seen significant regional shifts.

> **Exam tip**
>
> It is important to note the differences in levels of concentration of GHGs and to appreciate how they vary in their global warming potential. Methane is some 25 times more effective at trapping energy than carbon dioxide.

> **Knowledge check 3**
>
> What is the significance of the global warming potential (GWP) of different gases?

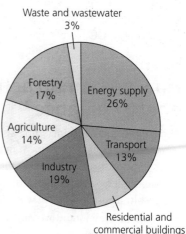

Figure 2 GHG emissions by source (source: IPPC (2007))

The top ten producers — made up of ACs such as the USA, Japan, South Korea and Canada, and Emerging and Developing Countries (EDCs) such as China, India, Iran and Saudi Arabia — still generate nearly 80% of all carbon dioxide emissions. ACs including the USA, Canada and Australia join EDCs such as the United Arab Emirates and Saudi Arabia in the top ten per person producers. China and India, in particular, have relatively low per person emissions.

Deforestation emissions are difficult to assess. However, when we take deforestation into account, countries such as Brazil and Indonesia are among the leading overall emitters.

How is the natural greenhouse effect enhanced?

Most GHGs occur naturally. They allow the incoming short-wave radiation from the sun to pass through the atmosphere relatively easily. However, GHGs absorb the long-wave radiation the Earth emits. They act like greenhouse glass, which reflects the sunlight, bouncing the energy around and allowing more heating to occur. Eventually the long-wave radiation makes its way back out to space. Without the natural greenhouse effect, the average temperature at the Earth's surface would be some 34°C lower.

As atmospheric GHG volume has increased, it has trapped more of the outgoing long-wave radiation. Attention is focused on the role of CO_2, as its increase has been so dramatic. But increases in water vapour and methane are also significant in enhancing the greenhouse effect.

Evidence from ice cores indicates that atmospheric CO_2 increases before overall global temperatures rise. Research has suggested that for the last four glacial–interglacial cycles, CO_2 levels increased up to 5,000 years before the ice sheets began to melt.

How do human activities influence the global energy balance?

The balance between incoming and outgoing energy determines the Earth's temperature. GHGs are critical in this balance, so any change in the atmosphere alters temperatures. Human activity has released considerable volumes of GHGs into the atmosphere over the past 200 years. The global energy balance has been altering as a result, leading to global warming:

- More outgoing long-wave radiation is absorbed and returned to the Earth's surface as back radiation, raising temperatures.
- Evaporation rates rise due to higher temperatures of both water and atmosphere. More latent heat transfers to the atmosphere in water vapour, which further alters the energy balance, as water vapour is a key GHG.
- Increasing temperatures lead to snow and ice melting. Less incoming solar radiation is reflected back to space but is absorbed by the land and sea, altering the energy balance.
- Land-use changes, such as deforestation, reduce albedo, increasing energy absorption and so altering the energy balance.

Exam tip

It is important to distinguish between total GHG emissions of a country and per person (capita) emissions. Both figures are significant, as future emission trends might see some substantial alterations to those figures.

Case study

For this part of the specification you are required to have case studies of **one AC** (e.g. Australia, Canada, France, Germany, Japan, the UK, the USA) and **one EDC** (e.g. Brazil, China, India, Indonesia, Mexico). Your case studies must illustrate:

- the contribution to anthropogenic GHG emissions from each country
- the principal activities contributing to GHGs, e.g. power generation, cement manufacture, transport
- the factors responsible for changes in GHG emissions, including the role of international protocols such as Kyoto or Paris

Why is there a debate over climate change?

Different agendas shape climate change debates

The very fact that there is debate indicates that countries, organisations and individuals have differing 'agendas' or perspectives regarding the nature and causes of climate change.

Historical background to the global warming debate

Historical evidence for global warming

The greenhouse effect was discovered in 1824. By the mid-nineteenth century, it was widely accepted that some gases had a role in trapping heat within the atmosphere.

In 1896, Swedish scientist Svante Arrhenius calculated that human activity could lead to global warming through adding CO_2 to the atmosphere. In the following decades, scientists believed human influences were minor compared with forcing mechanisms such as sunspots, the Earth's variable orbit and tilt, and ocean circulation. Many believed that oceans acted as vast sinks for excess CO_2, so there was little cause for concern.

In 1958, readings from both the Mauna Loa observatory, Hawaii, and Antarctica provided incontrovertible evidence that atmospheric CO_2 levels were steadily climbing. The 'Keeling curve' reveals that CO_2 has reached its highest level for some 700,000 years.

The global warming debate

Why was there a delay in recognising the impacts that GHG emissions and global warming could have? There are two reasons:

1 The nature of the data of global mean temperature
2 The lack of global environmental awareness

Exam tip

It is helpful to know something about the evolution of scientific thought regarding the Earth's atmosphere, giving context to recent changes in the climate change debate.

From 1940 to the mid-1970s, global temperature appeared to have a general downward curve. There was concern that the next ice age was beginning. From the late 1970s, significant advances were made in global climate modelling, and palaeoclimate research benefited from more reliable and accurate data, such as from deep-sea sediments. For example, it is now known that cooling in the 1960s and 1970s was due to the sunspot cycle and certain pollutants.

During the early 1980s, the global warming curve increased and by the end of the decade the rate of temperature increase was itself showing dramatic growth, as seen in the famous 'hockey stick' graph.

Although there are those that disagree, it is now the scientific consensus that global warming is occurring faster than at any previous time and that human activity is responsible for the increase. During the 1980s, a rise in environmental awareness occurred — the discovery of the hole in the ozone layer over Antarctica, the Chernobyl nuclear accident (1986) and the Exxon Valdez oil spill (1989) made headline news around the world. Global warming soon became a focus for concern, although a powerful denial lobby, involving both politics and business, still exists.

Paleoclimatology The study of past climates using, for example, ice cores, tree rings and organism remains found in sediment cores.

> **Exam tip**
>
> Construct a timeline to plot the key events in the evolution of the global warming debate. This will help you to understand how and why the debate has altered across the decades and bring accuracy to your exam answers.

The role of governments and international organisations

As climate change affects the entire planet, governments and international organisations have become crucial to the debate and consequent action or inaction.

The UN

Through the UN, thousands of people from a wide variety of disciplines come together to research and share findings. Some key landmarks have included:

- the 1988 Intergovernmental Panel on Climate Change (IPCC) set up by the UN Environmental Panel and World Meteorological Organisation
- the UN Framework Convention on Climate Change (UNFCCC), a 1992 international treaty signed by 41 countries at the Rio Earth Summit
- the 1998 Kyoto Protocol, which set legally binding targets to reduce GHG emissions — a total of 192 countries were involved but some (the USA, China) never ratified the treaty, while Low Income Developing Countries (LIDCs) were not required to cut GHG emissions

The EU

In regards to global warming, the EU is an environmental leader. The European Climate Change Programme (ECCP) sets targets for GHG reduction. It takes 1990 as the base year, setting legally binding targets of a 20% cut in GHG emissions, 20% of electricity from renewables and 20% improvement in energy efficiency by 2050. Progress towards targets varies across countries.

The cornerstone of this policy is the Emissions Trading Scheme (EU ETS). It operates a **cap** and trade system, and is the world's first major carbon market. It covers 45% of emissions, mainly CO_2 and N_2O. Each country has binding targets to reduce the remaining 55% from areas such as agriculture, housing, waste and transport.

The UK

The UK's Climate Change Act (2008) commits to a reduction in GHGs of at least 80% by 2050 compared with 1990 emission levels. The strategies include:

- short-term (5-year) carbon reduction targets
- increasing energy efficiency (improved building insulation, smart meters etc.)
- investing in low-carbon technologies (carbon capture and storage, renewables such as wind, biomass, solar)
- carbon taxes (on electricity generated by fossil fuels, especially coal, and vehicle CO_2 emissions etc.)

India

India did not ratify Kyoto, arguing that ACs were responsible for the increase in GHG emissions, and that EDCs and LIDCs should not have to pay. India argues that its annual per capita CO_2 emission of 1.7 tonnes is well below the global average of 5.0. Its priorities are reducing poverty and improving access to electricity among its population.

Nonetheless, India aims to lower GHG emissions by 20–25% by 2020, compared with its 2005 level. However, the commitment is voluntary and India's GHG emissions are expected to rise significantly over the next couple of decades.

The role of the media and interest groups

The media

Most people gain knowledge and understanding of issues such as climate change through access to popular media, and media interest in global warming has increased since the late 1980s. Particular political viewpoints tend to colour the media's reporting on global warming, for example right-leaning media tends to give greater prominence to sceptical opinions.

The media has also found it hard to understand and communicate scientific research methods. The fact that historic data are often recalibrated does not mean that those findings are invalid. For example, satellite data from the 1950s and 1960s have required reappraisal, to account for the contrasts in height above the Earth of different satellites.

Cap A 'cap' is set on the total amount of GHG to be emitted by energy users such as power stations, factories and airlines. Companies receive or buy emission allowances, which they can trade if they cut their emissions below their targets.

Exam tip

Research how countries across the development spectrum are responding to global warming to give your responses balance and authority.

Exam tip

So that you can evaluate points of view, try to decipher the political background of certain media. In this respect, you should treat websites in particular with caution.

Often reporters do not have the time nor space to provide more than generalisations about scientific research. Thus 'arguments' among researchers are presented as serious disagreements or 'proof' that the research is flawed.

Interest groups

Interest groups take various positions on climate change. The energy and mining industries have vested interests in maintaining the production of fossil fuels. They have considerable resources to spend on promoting themselves, including influencing decision makers such as politicians.

Other players support GHG reductions through their particular focus on aspects such as the environment or wildlife. Non-governmental organisations (NGOs) such as Christian Aid, Oxfam and Greenpeace operate on both national and international scales. Local media and organisations are also part of the climate change debate.

Exam tip

Investigate local media and interest groups' involvement in the climate change debate. They can be effective in supporting a discussion alongside the better-known national and international examples.

In what ways can humans respond to climate change?

Now that global warming is impacting an ever-increasing number of people, many are focusing on how humans can respond.

An effective response relies on knowing what the future holds

The response to the challenges created by global warming requires intensive preparation. To be effective, plans must be built upon high-quality information.

The importance of the carbon cycle

Carbon is ubiquitous on Earth, forming the basis of 95% of all known compounds. Stores, or **sinks** and flows of carbon, as well as processes involving carbon, are vital for life on Earth in a number of ways:

- Life is carbon based.
- Carbon stores such as carbonate rocks, fossil fuels and ocean sediments lock away carbon for millions of years — part of the slow carbon cycle.
- Carbon is stored in the atmosphere (e.g. CO_2), the oceans (dissolved CO_2), soils and living organisms — part of the fast carbon cycle.
- Green plants and phytoplankton convert CO_2 into carbohydrates, thereby beginning the flow of carbon.
- Decomposition and oxidation maintain the flow of carbon, e.g. CO_2 is returned back to the atmosphere.
- CO_2 and CH_4 are important GHGs, trapping long-wave radiation in the atmosphere.

Sink Anything that accumulates more of a particular substance than it releases, e.g. the oceans act as a sink for carbon dioxide.

Carbon is always on the move, and over long time scales the carbon cycle has a degree of balance. Recently, scientists have been concerned by the speed at which the balance is altering.

How feedback operates to affect climate change

A change in one part of the Earth's natural systems has consequences across the Earth. This change occurs due to feedback loops, both positive and negative.

Possible **positive** feedback loops include:

- increased evaporation → greenhouse effect enhanced → more warming → increased evaporation
- increased cloud cover (due to more water in atmosphere) → more outgoing long-wave radiation trapped → more warming → increased evaporation
- decreased albedo (due to ice and snowmelt) → more solar radiation absorbed at Earth's surface → more warming → decreased albedo
- increased ocean acidity (due to more CO_2 dissolved in oceans) → decrease in capacity of oceans to absorb CO_2 → more CO_2 in atmosphere → more warming
- permafrost thaws → release of vast amounts of CH_4 and CO_2 stored in previously frozen permafrost → these GHGs increase warming
- methane hydrates released from ocean sediments as oceans warm → methane a potent GHG → more warming
- decrease in forest cover due to increase in aridity in some areas → CO_2 released as trees decompose and less CO_2 stored in biomass → more warming

Possible **negative** feedback loops include:

- increased cloud cover (due to more water in atmosphere) → more incoming solar radiation reflected back into space → lower temperatures
- increased aerosols in atmosphere due to fossil burning → more incoming solar radiation reflected back into space → lower temperatures
- increased area of forest (due to warming of high latitudes and more CO_2) → more CO_2 stored in biomass

The outlook for GHG emissions and implications for global temperatures and sea levels

A range of predictions exists because the Earth–atmosphere–ocean systems have complex inter-relationships. Much also depends on what happens to the levels of GHG emissions. It is common in scientific investigation to have a range of outcomes based on various levels of probability.

Mean global temperature changes

IPCC predictions show significant temperature rises during this century. Depending on GHG emissions, temperatures will rise between 0.3°C and 4.8°C by 2100, assuming no natural forcing occurs, such as a major volcanic eruption. Given the lack of reduction in GHG emissions in some countries, the IPCC concludes that a rise of some 2°C compared with the early twenty-first century is very likely.

Mean global sea-level changes

The rise in sea level is predicted to occur at a faster rate through to 2100 than has been observed in the past. Between 1990 and 2010, it rose at just over 3.0 mm/year. As with temperature, there is a range of possible outcomes (0.28 m to just under

Positive feedback An automatic response to a change in a system or cycle that generates further change.

Negative feedback An automatic response to a change in a system or cycle that restores equilibrium or balance.

Permafrost Frozen ground that does not thaw out from one year to the next.

Methane hydrate Cage-like lattice of ice that contains trapped methane molecules. When warmed, the methane is released and then acts as a potent GHG.

Exam tip

Make sure you can give examples of both positive and negative feedback to provide substance to your responses.

1.0 m). The worst-case scenario, however, excludes the collapse of the West Antarctic ice sheet. If this were to happen, sea-level rise would be well above 1 m.

Climate change impacts are global and dynamic

The implications for people and the environment

Climate change will increase the intensity and occurrence of existing risks as well as bringing new risks for human communities and the natural environment.

Ecosystems

The changes that global warming brings will have feedback consequences for the abiotic (non-living) and biotic (living) components in ecosystems.

Marine ecosystems

The rise in SSTs is likely to be particularly intense in tropical and subtropical regions. Coral ecosystems are at risk due to **coral bleaching**, which can lead to coral mortality and the loss of one of the most diverse and productive ecosystems on Earth (an 80% loss has been recorded in the Seychelles and the Maldives).

Sea ice loss around the Arctic Ocean has led to substantial reductions in algae and plankton, which are the basis of the ecosystem. Fragmentation and loss of sea ice affects all mammals living in the region. Walrus, seals and polar bears all rely on being able to move across the ice or use it as a platform on which to rest, give birth and raise young.

There has also been a 30% increase in the acidity of oceans and the forecast is for acidity to double by 2100. This reduces the ability of shell-building organisms to accumulate calcium carbonate. As many of these creatures form the lower **trophic levels** of ecosystems, feedback means that organisms throughout the oceans are threatened.

Terrestrial ecosystems

The tundra is the treeless region in the sub-Arctic and at high altitudes, which has a short growing season (50–60 days) and low winter temperatures (average −34°C). Much of the tundra exists as permafrost. But warming is thawing the permafrost, expanding wetland areas, and lengthening growing and breeding seasons, and the tree line is migrating north.

Migration patterns of birds, especially waders and wild fowl (ducks, geese) and animals such as caribou, are being altered. Some predatory species (snowy owl, Arctic fox) tend to hunt in the open so the growth of trees and shrubs affects them negatively.

Areas of tropical cloud forests are also shrinking. Species such as the mountain gorilla in Central Africa and many amphibians cannot migrate in response to climate change as either they live in isolated populations and/or humans occupy potential alternative habitats.

Exam tip

The IPCC maintains a detailed website that is constantly updated. Look for the 'Synthesis' parts of its main reports, as these summarise the IPCC's conclusions.

Coral bleaching The loss of algae living within the coral due to higher water temperature. The algae gives the coral its colours.

Trophic level Level at which energy in the form of food is transferred from one organism to another as part of food chains and webs.

Knowledge check 4

Why has ocean acidification the potential to have a very serious impact on both marine ecosystems and human populations?

Human health

Increased global temperatures will raise levels of ill-health and mortality. The most serious implications arise from changes in access to fresh drinking water. Altering river flows and the re-filling of groundwater stores affect the supply of water, in particular for the growing numbers of people living in urban areas in EDCs and LIDCs. Rising sea levels threaten coastal **aquifers** with contamination from sea water. More intense rainfall increases flood risk and this can lead to water pollution via animal and human waste. Water-borne diseases such as diarrhoea and cholera can quickly spread as a result.

Rising temperatures allow insects to extend their habitat. If an insect is a **vector** for a disease, that poses increased risks to human health. Dengue fever is spreading northwards through the USA as the Aedes mosquito extends its range, while southern Europe is at risk from the spread of malaria.

As temperatures rise, the risk of food contamination also increases, as bacteria thrive under warm conditions. To combat this threat, more power is required so that refrigeration is available. This is a serious concern in many LIDCs.

Aquifer A rock capable of storing water, such as chalk.

Vector Organisms that can transmit disease, e.g. mosquito → malaria.

Extreme weather

Links between global warming and extreme weather are complex and the subject of much ongoing research. It is important to recognise that extreme weather is by its very nature variable, so establishing clear trends is difficult.

- **Tropical storms:** there is evidence for an increase in the number and intensity. Rising SSTs fuel tropical storms. Higher rates of evaporation generate more water vapour, which carries vast amounts of latent energy into the atmosphere that, when released during condensation, powers tropical storms.
- **Monsoon:** temperature contrasts between the oceans and continents drive the monsoon. Climate models predict that this contrast will widen when, during summer, continents warm more than oceans; reduced snow cover on the Tibetan plateau increases temperature differences between land and sea, and warmer seas and air increase evaporation so that the atmosphere carries more water vapour. The predictions are for increases in average rainfall, the number of days with very heavy rain, and in year-on-year monsoon variability.
- **Heat waves and droughts:** heat waves, such as that seen in 2003 in Europe (with an estimated 35,000 dead) are likely to be average summer temperatures by 2050. As precipitation becomes more unpredictable, drought will increase.
- **Mid-latitude storms:** there is evidence that regions such as north-west Europe have become stormier over the past 50 years, with more intense rainfall.

People and environments: vulnerability

People

Physical, social, economic and political factors all influence vulnerability. Each person, family, community and country has a range of climate change effects it can cope with, usually based on past experiences and predictable risks. Occasionally an extreme event occurs which can lead to disaster, such as Hurricane Katrina.

- **Subsistence farmers:** as the climate moves to new averages in temperature and precipitation, people will become more vulnerable unless their ability to cope increases. In general, ACs and some EDCs are likely to have the resources to cope, whereas vulnerability among the LIDCs is already an issue and will become ever more serious. In particular, subsistence or semi-subsistence farmers are at great risk, as most rely on **direct rainfall** for their crops and livestock to survive.

- **Marginal locations:** those living in **marginal locations** are at greatest risk as they already face great variability in weather and have few resources to fall back on when an extreme event occurs. Locations in ACs, such as the prairies in the USA or steppes in Russia (EDC), are at risk of becoming more marginal for agriculture if rainfall becomes more erratic.
- **Coastal regions:** those living in coastal regions or on small, low-lying islands are vulnerable to rising sea level. This is true for locations across the development continuum, although ACs possess greater resources (such as sea defences) for reducing vulnerability.
- **The Arctic:** indigenous peoples are extremely vulnerable, as sea ice thins and its area shrinks. Traditional ways of life are increasingly threatened.
- **Urban dwellers:** the built environment absorbs, stores and releases heat energy and wind speeds are lower, making urban dwellers, in particular the elderly, the young and the chronically ill, more vulnerable.

Environments

Climate change will affect **biodiversity** in a wide range of ecosystems such as deserts, mountains, tundra, tropical rainforest and estuaries (see Ecosystems, page 18).

Mitigation and adaptation strategies for dealing with the risks from climate change

The most sensible approach to dealing with risks from climate change is to cut carbon dioxide emissions. **Mitigation strategies** aim to reduce GHG emissions in order to restrict global warming. **Adaptation strategies** aim to offer greater protection to those people and environments already facing risks from climate change.

Mitigation strategies to cut GHG emissions

Energy efficiency and conservation

Mitigation strategies in energy efficiency and conservation include:

- improving vehicular fuel efficiency
- encouraging the use of public transport
- stricter regulations on building insulation
- improving the efficiency of coal-fired electricity plants (currently about 32% up to 60%)

Fuel shifts and low-carbon energy sources

Mitigation strategies in low-carbon energy sources include switching:

- coal for gas in power stations (less CO_2 emitted)
- nuclear for fossil fuel in power generation
- wind power for fossil fuel in power generation
- photo-voltaic (solar) for fossil fuel in power generation

Direct rainfall That which falls onto the fields (excludes water stored in reservoirs, or river/lake water that farmers use in irrigation).

Marginal locations Those regions that sit at the edge or limit of people's ability to survive (e.g. Sahel, north Africa). Often it is the unreliability of rainfall that makes a location marginal.

Biodiversity Refers to the degree of variability within species, between species and between ecosystems. A high degree of biodiversity is considered to be important for a healthy environment in any particular location.

Exam tip

There are pros and cons to each of these mitigation strategies. Make sure you are able to offer these in any discussion of possible ways to cut GHG emissions.

- biomass fuel for fossil fuel in power generation
- tidal and wave power generation for fossil fuels (not commercially viable at present)
- geothermal and ground-sourced heat pumps.

Carbon capture and storage (CCS)

- introduce CCS at coal-fired power stations

Forestry

- decrease deforestation rates in tropical regions and increase afforestation in all areas of the world

Geoengineering

Mitigation strategies in geoengineering include:

- fertilising oceans with iron to stimulate phytoplankton growth, which absorbs carbon dioxide so that it is stored in deep ocean sediments when the organisms die
- mirrors in space to deflect incoming solar radiation
- injecting aerosols into the atmosphere to reflect incoming solar radiation
- producing artificial plastic trees that absorb CO_2, which can then be stored

Geoengineering strategies have the major disadvantage that we simply do not know what consequences they might have. In addition, costs and practical considerations suggest that we would do better to focus on reducing our production of carbon.

Adaptation strategies to reduce people's vulnerability

For each one of the following strategies, it is important to consider the contrasting abilities of societies across the development continuum to implement them.

Retreat

Retreat may involve:

- **managed realignment**, relocating buildings and agriculture, and creating new habitats
- abandoning low-lying coastal locations altogether
- zoning vulnerable floodplain locations to prevent building

Accommodate

Accommodation may involve the following strategies.

- **Agriculture:** growing more drought-resistant crops, possibly genetically modified (GM), **zero tillage**, crop rotation, planting hedges to reduce wind speeds, which in turn reduces evapotranspiration
- **Water supply and use:** increasing the collection, storage and recycling of water e.g. **grey water**, more efficient use of water in machines, reducing losses through better detection and swift repair of leaking pipes, desalinisation, inter-basin transfer

Protect

Protection may involve:

- hard engineering, e.g. flood defences such as storm surge barriers, dams, levees
- soft engineering, e.g. encouraging mangrove, saltmarsh and sand dune growth to absorb wave energy, and afforestation in upland areas to store water and slow its path through the basin

CCS This involves removing CO_2 during industrial processes and power generation and piping it underground to where oil and gas have been extracted.

Managed realignment Approach whereby present-day coastline is no longer protected and land is allowed to flood. Further inland, new defences are constructed and often, between these and the sea, saltmarsh is encouraged.

Zero tillage Growing crops without disturbing the soil through activities such as ploughing, which reduces soil moisture loss (also known as 'no till').

Grey water Water that has already been used and is still suitable for flushing toilets, watering gardens etc.

- health measures to protect against migrating insect vectors of disease, e.g. reduce breeding sites, use pesticides and increase access to vaccination
- health measures to protect against extreme heat, e.g. increased use of air conditioning

Table 1 shows ways in which we can adapt buildings, transport and urban areas to counter the impacts of climate change.

Table 1 Climate change adaptations for building, transport and urban areas

Type of impact	Buildings	Transport	Urban areas
High temperatures	Air-conditioning, shade windows, green roofs assist cooling	Air conditioning, design greater resilience into technology, e.g. upgrade steel in rails to prevent buckling and tarmac able to withstand heat	Green spaces help cool locations, more sub-stations and cable routes to ensure system keeps working under high demand
Droughts	Rainwater harvesting off roofs, grey-water systems, deeper foundations as soil shrinks	N/A	Capture, store and recycle grey water, encourage and legislate for efficient use of water
Floods	Use basements and ground floors for car parks, stilt houses, water-proof electrical supply, roof gardens to reduce runoff, floating buildings	Raise vulnerable equipment, e.g. electricals above flood risk, improve drainage e.g. culvert size, plan for use of alternative routes	Reduce impermeable surface area, increase vegetation area, redesign drainage systems e.g. allow water to soak into aquifers, land-use planning and zoning

Economies

Most ACs have resources to cope with climate change. Some of the EDCs, such as Chile, Latvia and Poland, are likely to cope if their economic resources allow.

The LIDCs are most vulnerable as they are already at risk from events such as floods and droughts. Their economies tend to rely on climate-sensitive activities, such as agriculture and tourism. Countries like Bangladesh and Egypt have significant populations living on river deltas. Such locations are increasingly prone to coastal and river flooding.

Knowledge check 5

What is the difference between mitigation and adaptation in the context of climate change?

Case studies

For this part of the specification you are required to have case studies of **two countries** at contrasting stages of economic development. Examples include:

- Australia, Canada, Germany, Japan, the USA (ACs)
- Brazil, Egypt, Indonesia, Russia (EDCs)
- Bangladesh, Chad, Maldives, Vietnam (LIDCs)

Each case study should illustrate:

- current socioeconomic and environmental impacts due to climate change
- the opportunities and threats these impacts present
- the technological, socioeconomic and political challenges associated with effective mitigation and adaptation strategies

Can international responses to climate change ever work?

Effective implementation depends on policies and cooperation at all scales

If we are to combat climate change, there needs to be a system of international responses, with countries across the development spectrum working together to effect change.

Geopolitics associated with human responses to climate change

The IPCC

Established in 1988, the Intergovernmental Panel on Climate Change (IPCC) brings together all the key research on climate change and produces a general agreement among the researchers. The purpose of the IPCC is to provide ongoing objective assessments of:

- the science behind climate change
- the environmental and socioeconomic impacts (positive and negative) of climate change and options for adapting to it
- policy options as regards limiting GHG emissions

As so many leading figures covering a broad range of expertise from all round the world contribute to the IPCC, and because its reports are subject to intensive review, the IPCC is recognised as the most authoritative voice on climate change. In 2008, the IPCC was awarded the Nobel Peace Prize, an indication of the global importance of its work.

The success of international directives

The Kyoto Protocol

Most of the world's ACs support the Kyoto Protocol, with the notable exception of the USA. Targets for emission reduction were reduced to make sure that Australia, Canada and Japan signed. EDCs, such as China and India, and LIDCs were exempt. Issues with Kyoto include the following.

- **Reductions in GHG emissions did not go far enough:** given the countries not signed up, targets needed to be much more ambitious.
- **No enforcement:** like many international directives, how can sanctions be applied to countries that sign up and then do not comply?
- **No USA:** it is the second-highest national emitter and one of the highest per capita. Successive US governments have either been opposed to cuts or faced very strong domestic opposition.
- **Focus on the largest emitters:** fewer than 20 countries emit some 80% of global GHG emissions. It's suggested that trying to obtain an agreement covering over 150 countries is not that helpful, or realistic.
- **National vs sector emissions:** it may be more suitable to get agreements by sector in order to limit emissions per tonne of product, such as steel or cement.

Carbon trading and carbon credits

Kyoto included the idea of **carbon trading** and **carbon credits**. Its basis is that if the polluter pays for pollution, they would then find ways of reducing it.

Some success has been achieved through cap and trade (see page 15). In 2005, the EU Emissions Trading System (EUETS) began limiting emissions from more than 11,000 heavy energy-using installations (power stations and factories, such as oil refineries, steel, cement) and airlines flying within Europe. This now covers about 45% of the EU's GHG emissions and aims to achieve a 20% reduction by 2020 compared with the 1990 level, rising to 43% by 2030.

The EUETS has achieved significant reductions in emissions — some 24% between 2005 and 2015. However, the trend in reduction has decreased and further measures will be needed to meet the 2030 target. Much of the trading revenue goes into research, renewable energy and improving energy efficiency projects.

National and subnational (regional) policies

Denmark is phasing out fossil fuel use, and investing in wind and solar power and public transport. It is also introducing stricter controls on agricultural methane emissions. Copenhagen has introduced adaptation policies such as storm barriers to protect against the increasing risk of storm surges.

California, USA has also adopted a range of radical mitigation policies:

- a phased programme to eliminate fossil fuel power stations, short-term switch to gas
- stringent laws regarding vehicle emissions (target of 15% electric/hydrogen fuel/compressed natural gas cars by 2020)
- target of 33% electricity from renewables by 2020
- cap and trade scheme covering 85% of carbon polluters in the state — second-largest after the EUETS

California's adaptation policies include water management, which in a semi-arid region with such a high demand is a serious concern. Parts of southern California also face risks from wildfires, and flooding and erosion following intense rainfall.

Carbon trading Allows businesses to sell any of the **carbon credits** (quota) they do not use. Other businesses can purchase these to offset against any carbon they produce in excess of their own quota.

Exam tip

Use the internet to keep up to date with facts and figures relating to schemes such as the EUETS.

Summary

- The Earth's climate is dynamic, changing throughout geological time.
- Various natural forcing factors operate across long time scales (geological).
- There are several methods used to reconstruct past climates.
- Humans have influenced the climate system to such an extent that the current geological epoch is known as the Anthropocene.
- Human activity has added GHGs to the atmosphere, enhancing the natural greenhouse effect.
- Different agendas shape the debate over climate change.
- Climate modelling is important in planning responses to climate change.
- Impacts of climate change are dynamic and global.
- Mitigation and adaptation strategies reduce and manage risks.
- Geopolitics is a key influence on responses to climate change.

■ Disease dilemmas

What are the global patterns of disease and can factors be identified that determine these?

Classification of diseases and their geographical distribution

Disease classification

Global patterns of disease are more easily identified and understood if we distinguish between different types of disease (see Table 2).

Table 2 Disease classifications

Disease classification	Mode of transmission/cause	Examples
Infectious	A disease caused by pathogenic microorganisms such as bacteria, viruses, parasites or fungi. These can be spread directly or indirectly from one person to another.	Influenza, pneumonia, malaria, TB, HIV/AIDS, poliomyelitis, yellow fever, measles, cholera, Zika virus
Non-infectious	A non-communicable disease, not caused by pathogens. Diet, environment, lifestyle, age, gender and inherited genetics affect the risk.	Asthma, diabetes, cancer, stroke, cystic fibrosis
Communicable	A disease that spreads from host to host. Pathogens passed from person to person or from animal to person can cause it.	Common cold, Ebola, rabies, influenza, HIV/AIDS, hepatitis, poliomyelitis, TB
Non-communicable	A disease that cannot spread between people because it is non-infectious or non-contagious. Lack of physical activity, smoking or poor diet, exposure to air pollution, genetic defects, age and gender may increase the risk.	Diabetes (types 1 and 2), coronary heart disease, osteoporosis, Alzheimer's, asthma, lung cancer, leukaemia, skin cancer
Contagious	An infectious disease caused by bacteria and spread by direct physical contact or indirect contact between people.	Common cold, influenza, HIV/AIDS, typhoid, plague, rubella, TB
Non-contagious	Diseases not due to disease-causing organisms but caused by genetics, diet, lifestyle or environment.	Genetic diseases such as sickle-cell disease or cystic fibrosis, and cardiovascular disease, skin cancer

It is also important to understand the following terms in relation to global patterns of disease.

- **Endemic diseases:** these exist permanently in a geographical area or in a specific human group. The disease is not necessarily present at a high level of occurrence but it can always be found in that population. For example, malaria is endemic in many parts of Africa.
- **Epidemics:** disease outbreaks that spread quickly through the population of a geographical area affecting a large number of people at the same time (e.g. Ebola epidemic in west Africa, 2013).

- **Pandemics:** epidemic disease outbreaks that spread worldwide, for example when a new virus emerges for which most people do not have pre-existing immunity (e.g. H1N1 flu virus, 2009–10).

Global distribution of diseases

Malaria

Malaria risk is greatest within the tropics — 90% of deaths in 2015 were in Africa, with greatest numbers in Nigeria and the Democratic Republic of the Congo (DRC).

Malaria is an **infectious** but non-contagious disease. It is caused when the female Anopheles mosquito (**vector**) takes a blood meal from an infected person and then injects the parasite (plasmodium) when taking a blood meal from an uninfected person. Young children, pregnant women and non-immune tourists are particularly vulnerable populations.

Global distribution is influenced by climatic factors, especially temperature but also humidity and rainfall. The Anopheles mosquito thrives in warm, humid environments where there is stagnant water, in which it lays its larvae. According to the World Health Organization (WHO), malaria is **endemic** in 95 countries. Transmission is all year round close to the equator, and seasonal (towards the end of the rainy season) further from the equator.

Malaria transmission cannot occur in all parts of countries where it is endemic. Risk is much lower in areas of high altitude, aridity or during a cold season, and where there has been successful intervention. This is because the Plasmodium (P.) falciparum parasite, which causes the most severe malaria, cannot complete its cycle of growth in the female Anopheles mosquito in these conditions.

Beyond the tropics, areas of risk do exist but transmission is more limited in areas such as southwest Mexico and eastern China. In temperate areas, such as western Europe and North America, where there are high levels of investment in public health, risk of transmission is very low. Over time, in the LIDCs of central Africa, use of insecticides, drainage of breeding areas, mosquito nets and education have lowered the risk. Nevertheless malaria is difficult to control, especially where human factors such as poor sanitation and presence of large, high-density populations contribute to the risk. Further statistical evidence and information on factors affecting global distribution of malaria are available in the WHO annual *World Malaria Reports*: www.who.int/malaria/publications/world_malaria_report/en/

HIV

Human immunodeficiency virus (HIV) is a **communicable disease** that is infectious and contagious. Globally, there is significant variation in the prevalence of HIV, but a particularly high proportion of the infected adult population is found in sub-Saharan Africa (e.g. Zambia and South Africa). HIV could lead to development of AIDS, its most advanced stage, and it reduces natural immunity to other viruses. According to the WHO, 1.1 million people died from HIV-related causes in 2015 (see Table 3).

Exam tip

Accurate use of terminology in exam answers helps to demonstrate knowledge and understanding.

Infectious disease A disease spread by parasites, bacteria, viruses or fungi.

Vector Living organisms such as mosquitoes, ticks and some freshwater aquatic snails that can transmit infectious diseases between humans or from animals to humans.

Endemic disease A disease that exists permanently in a geographical area or human group.

Communicable disease An infectious disease which is transmissible between people through a variety of ways, including contact with blood and body fluids; breathing in an airborne virus; being bitten by an insect; and contact with a contaminated surface.

Table 3 Estimated prevalence of HIV/AIDS, 2015

WHO region	Adults and children estimated to be living with HIV (millions)	Deaths due to HIV/AIDS
Africa	25.5	800,000
Southeast Asia	3.5	130,000
The Americas	3.4	62,000
Europe	2.5	56,000
Western Pacific	1.4	44,000
Eastern Mediterranean	0.3	15,000

Transmission of the HIV virus is by body fluids, such as blood, breast milk and semen, from infected persons. The risk of infection increases through unprotected sex, having another sexually transmitted disease such as syphilis or herpes, sharing contaminated needles when injecting drugs and receiving unsafe blood transfusions. Lower percentages of HIV-affected adults tend to prevail in more developed countries, where research, drugs and education programmes are readily available. In many LIDCs, higher percentages of HIV-infected adults are explained by:

- limited funding and availability of drugs
- insufficient numbers of trained medical staff, especially in rural areas
- high birth rates among infected women
- high levels of illiteracy

However, progress is being made in some LIDCs, such as Malawi, which includes:

- self-testing for HIV where testing may not otherwise be available
- anti-retroviral treatment
- elimination of mother-to-child transmission of HIV

Information on factors affecting global distribution of HIV is available in the WHO *Global Health Sector Strategy on HIV, 2016–21*: www.who.int/hiv/strategy2016-2021/ghss-hiv/en/.

Tuberculosis

According to the WHO, 1.8 million people died from tuberculosis (TB) in 2015 — 95% of these deaths were in low- and middle-income developing countries. Estimates of new cases are greatest in sub-Saharan Africa, especially south of the equator (see Figure 3).

TB is an infectious and highly **contagious** communicable disease. It is a bacterial infection spread by transmission of mycobacterium tuberculosis from person to person through the air (via inhalation of droplets from coughs and sneezes of infected people) and it typically affects the lungs.

Contagious disease A disease spread by contact or indirect contact between people.

Incidence of TB is worldwide, although 60% of deaths in 2015 were in six countries: China, India, Indonesia, Nigeria, Pakistan and South Africa. Risk factors that affect its distribution include living conditions in poor communities where there is overcrowding and there are high-density populations and poorly ventilated houses. Limited access to health services is a significant adverse factor, especially in areas affected by civil unrest or war. Incidence of TB is also higher where an individual's immune system is compromised because they are living with HIV, **malnutrition** or diabetes — there were 0.4 million TB deaths among people with HIV in 2015.

Malnutrition The shortage of proteins and essential vitamins caused by an unbalanced diet.

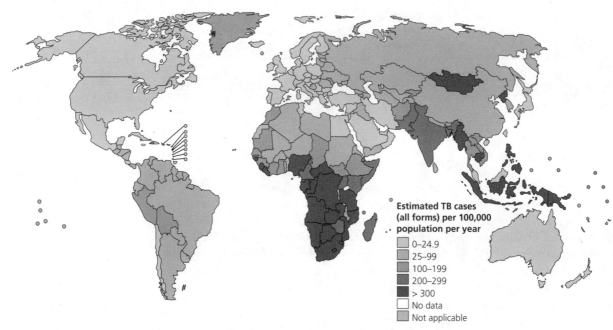

Estimated TB cases
(all forms) per 100,000
population per year

- 0–24.9
- 25–99
- 100–199
- 200–299
- > 300
- No data
- Not applicable

Figure 3 Estimated new TB cases (2015)

Ending the TB epidemic by 2030 is one of the UN health targets in Sustainable Development Goal 3 (SDG3). Details of the research into geographical patterns, trends and strategies are available in the WHO *Global Tuberculosis Reports*, which are updated annually at www.who.int/tb/publications/global_report/en/.

Diabetes

Diabetes is a non-communicable disease that can lead to heart, blood vessel, eye, kidney and nerve damage. Type-1 diabetes is the result of the body's deficiency in insulin production. This can develop at any age but often begins in childhood. It is thought to be genetic and not related to diet or lifestyle. Type-2 diabetes is the result of the body's ineffective use of insulin. Approximately 90% of people with diabetes have type-2. The main risk factors for type-2 diabetes are excess body weight, physical inactivity, age (diabetes may develop in later life), smoking and poor diet.

In 2014, there were 422 million people living with diabetes. The disease is widespread in all WHO regions, but there is significant variation. Prevalence is high in North America and in east and southeast Asia, and lower in most of Central Africa, Central America and Nordic countries. In the UK there are thought to be over 4 million people affected. By number of prescriptions for the disease per person, highest incidence occurs in the London boroughs of Newham and Tower Hamlets, and in coastal Lincolnshire.

The number of type-2 diabetes cases is rising more rapidly in low- and middle-income countries than in developed countries. This is the result of an increase in overweight and obesity in developing countries, in both children and adults.

Many people remain undiagnosed, but the WHO is improving research and surveillance in order to prevent and manage diabetes, and to make recommendations for governments and civil society organisations (CSOs). Tackling obesity through

Exam tip

The ability to analyse statistical maps/diagrams of spatial patterns of disease and changes over time is an important skill.

education, establishing good eating habits and encouraging physical activity from an early age is important. Further data, factors and strategies, plus links to country profiles, are available in the WHO *Global Report on Diabetes*: www.who.int/diabetes/global-report/en/.

Cardiovascular disease

Cardiovascular disease (CVD) includes a range of disorders of the heart and blood vessels. These are non-communicable diseases — they cannot be passed from person to person — and include high blood pressure, heart attack and stroke. Globally, CVDs account for 17.5 million deaths per year. CVDs are prevalent in all WHO regions, often among older age groups, but there are also many premature deaths before the age of 70. Low- and middle-income countries are disproportionately affected, having over 80% of all CVD deaths. Mortality rates are particularly high in the middle east, eastern Europe, south Asia and sub-Saharan Africa.

CVD risk is increased through a combination of lifestyle factors such as smoking, harmful use of alcohol, unhealthy diet and physical inactivity. In addition, underlying causes include population ageing, poverty and hereditary factors. Rapid, unplanned urbanisation can have negative socioeconomic effects, exposing children to crowded living, air and water pollution, inadequate sanitation and tobacco, alcohol and fast food. There is higher incidence where there is poor access to medical care.

Low-income countries and groups have less capacity to control and prevent CVDs. Governments are encouraged to participate in the WHO's Global Heart Initiative. Global organisations such as the World Heart Federation and the World Stroke Organization are increasingly effective in reducing death rates through education.

Further statistical evidence, specific causes and interventions are available in the WHO *Global Status Report on NCDs, 2014* (www.who.int/nmh/publications/ncd-status-report-2014/en/) and the *Global Action Plan for the prevention and control of NCDs, 2013–20* (www.who.int/nmh/events/ncd_action_plan/en/).

Disease diffusion, including the Hägerstrand model

Types of diffusion

Diffusion is the process by which a particular disease spreads outwards from its geographical source. Hägerstrand's model, based on the spread of agricultural innovation in central Sweden, can be applied to spatial patterns of disease diffusion. The model identifies four stages of the diffusion process.

1 **Primary stage:** there is strong contrast in disease incidence between the area of outbreak and more remote areas.
2 **Diffusion stage:** diffusion is centrifugal — new centres of disease outbreak occur at distance from the source and this reduces the spatial contrasts of the primary stage.
3 **Condensing stage:** the number of new cases is more equal in all locations, irrespective of distance from the source.
4 **Saturation stage:** diffusion decelerates as the incidence of the disease reaches its peak.

Processes of disease diffusion include the following.

- **Expansion diffusion:** the disease spreads from one place to another, forming new areas of prevalence, and it remains in the area of outbreak, possibly even intensifying there.
- **Relocation diffusion:** the disease moves to new areas but does not remain in the area of outbreak.
- Contagious diffusion: distance is a factor in the spread of contagious diseases — there is greater chance of a contagious disease being passed on to people living near the source than those further away.
- Hierarchical diffusion: a disease may be transmitted, usually down the urban hierarchy, from larger, more accessible towns to more remote, rural villages.

The spatial patterns and stages of the model are illustrated by the Ebola epidemic in west Africa. Ebola is an infectious and contagious communicable disease. The virus is transmitted by contact with infected animals or body fluids of infected persons. A combination of environmental, social, economic and political factors influences the pattern and rate of diffusion.

Barriers to diffusion

Physical barriers to the spread of disease include high mountain ranges, large maritime areas, extensive areas of aridity and climatic conditions such as the extent of tropical conditions in a continental land mass. Distance is also an important factor.

Socioeconomic barriers include national government and international organisation strategies to tackle communicable diseases, such as medical health checks at international borders and airports, quarantine, vaccination and health education programmes.

The relationship between physical factors and prevalence of disease

Global patterns of temperature, precipitation, relief and water sources affect patterns of disease

Patterns of disease are affected by the following.

- **Climate:** temperature, rainfall and humidity influence habitats of disease vectors. For example, the Anopheles mosquito, which transmits malaria, and the tsetse fly, which transmits sleeping sickness, are both endemic in tropical conditions.
- **Relief:** the influence of altitude on temperature and rainfall regimes affects vector habitats. For example, in Ethiopia at over 2000 m it is usually too cold for P. falciparum to develop in the mosquito vector.
- **Water sources:** stagnant water affects prevalence of water-borne diseases, such as Guinea worm disease.

Contagious diffusion The process by which a disease spreads through direct contact with a carrier.

Hierarchical diffusion The process by which a disease spreads through a structured order of places.

Knowledge check 6

What is meant by 'disease diffusion'?

Exam tip

Physical factors have a strong influence on disease patterns. Make sure you know the specific impact of different types of physical factors on disease patterns, including their influence on disease vectors.

Physical factors can influence vectors of disease

The WHO states that:

> Vectors are living organisms that can transmit infectious diseases between humans or from animals to humans. Many of these vectors are bloodsucking insects...

Examples include mosquitos, tsetse flies, ticks, fleas and some freshwater aquatic snails. Most of these vectors transmit pathogens via their blood meals. Vector-borne diseases include dengue fever, West Nile virus (WNV), African sleeping sickness, Zika virus, Chagas disease and malaria.

Globally, the vector-borne disease causing most deaths is malaria. Malaria depends on three physical factors.

1 **Rainfall:** female Anopheles mosquitos lay their eggs in water — these hatch into larvae. The abundance of aquatic habitats, ideally unpolluted fresh water, depends on collection of water that is not flowing, such as puddles towards the end of, or just after, the rainy season.

2 **Temperature:** where average temperatures are between 18°C and 40°C the mosquito takes more blood meals and increases the number of eggs laid, increasing the number of vectors. The larvae develop faster at higher temperatures and so the parasite has more time to complete its life cycle inside the mosquito.

3 **Relative humidity (RH):** where average monthly RH is over 60%, often increased by vegetation growth, the mosquito has a better chance of survival and becomes more active.

How seasonal variations influence disease outbreaks

Seasonality of temperature and rainfall regimes influences the prevalence of vector-borne infectious diseases. These factors affect time available for vectors and parasites to complete their life cycles and the availability of suitable aquatic habitats. Examples include the following:

- **Malaria:** outbreaks are closely linked to seasonal changes in rainfall. Relatively sudden transmission of P. falciparum occurs where seasonality is most marked, such as in tropical areas that are further from the equator in Africa or following monsoonal rainfall in Asia.
- **Sleeping sickness:** in the woodland savanna of west and central Africa, outbreaks occur in the wet season when the tsetse fly vector can live longer. There is significant regression of both vector and parasite in the dry season.

In temperate latitudes of Europe and North America, influenza epidemics tend to peak in the winter. Transmission of the influenza virus is more efficient at lower temperatures and when relative humidity is low. These conditions occur more often during the winter season.

Climate change provides conditions for emerging infectious diseases

Emerging infectious diseases (EIDs) are new diseases that have emerged in the last 20 years. Either they have not occurred in human populations before or they have occurred previously but affected only small numbers in isolated places. Many are spread by mosquito bites, including malaria, yellow fever, dengue fever, Zika virus and WNV. Outbreaks and diffusion of these diseases can result from changes in climatic conditions in which vectors survive and develop.

Diseases of concern for the WHO and health authorities in the Americas include Zika virus, WNV and Lyme disease. Global warming has had the effect of extending the geographical areas in which these vector-borne diseases are developing. WNV and Lyme disease are expected to spread northwards within the USA and possibly into Canada during the twenty-first century.

Some EIDs have re-emerged in areas where conditions have been made favourable to their transmission. Public health controls had eradicated the mosquito population that spread dengue fever in Florida, but increased temperatures have contributed to its re-emergence.

Shorter-term climate changes include the impact of El Niño Southern Oscillation on rainfall and temperatures. Effects such as higher temperatures, heavier rainfall, greater frequency of tropical cyclones and flooding have been persistent enough in some regions to create conditions for transmission of diseases such as malaria, cholera or dengue fever.

The conditions for zoonotic infectious diseases

Zoonotic diseases (**zoonoses**) are caused by bacteria (salmonella, anthrax, E. coli, leptospirosis), viruses (rabies, avian flu, Ebola), parasites and fungi. Zoonotic diseases are transmitted to humans from disease reservoirs of non-human species through:

- direct contact with animals, such as dog or bat bite (rabies)
- insect vector bites, such as mosquitos (yellow fever)
- contaminated food or water — a wide range of diseases, such as typhoid, dysentery, cholera and gastroenteritis, can be transmitted to humans by organisms that thrive in water contaminated by human faeces

Rabies is an infectious viral disease causing about 60,000 deaths annually worldwide. A high proportion of these occur in Africa and Asia, children being frequent victims. Incidence is highest in remote rural communities where vaccines are not readily available or accessible.

Emerging disease
A disease that has appeared in a population for the first time, or that may have existed previously but is rapidly increasing in incidence or geographic range (WHO).

Zoonoses Diseases or infections naturally transmitted between vertebrate animals and humans.

Knowledge check 7

What are zoonotic diseases?

Natural hazards can influence the outbreak and spread of disease

This part of the specification requires a case study of any **one country** to illustrate how a natural hazard can influence the outbreak and spread of a named disease. Recent examples include Bangladesh, where flooding in 2007 was linked to epidemics of diarrhoea and water-borne diseases such as typhoid, and Haiti, where a cholera outbreak followed the 2010 earthquake. Your case study must illustrate:

- the geographical area affected by the hazard and its influence on risk and outbreak of the chosen disease
- environmental and human factors that have affected the spread of the disease — these could include climate, water supply, population density and poverty levels
- the disease's short- and long-term effects on local residents
- national and international strategies to minimise impacts of the disease

Exam tip

In your answers, explanations should demonstrate secure understanding of all the factors (economic, social, political and environmental) influencing spatial patterns of diseases and their diffusion over time.

Is there a link between disease and levels of economic development?

As countries develop, the ratio of communicable diseases to non-communicable diseases changes

The WHO states that:

> As life expectancy increases, the major causes of death and disability in general shift from communicable, maternal and perinatal causes to chronic, non-communicable ones.

Epidemiological transition

In the UK, the main causes of death in the nineteenth century were infectious diseases, especially smallpox, cholera, typhoid and diarrhoea. Vaccination, more effective public health policy, clean water and improved nutrition, sanitation and housing have led to significant decline in deaths from these diseases — in the twenty-first century, such diseases cause a very low percentage of all deaths.

The main causes of death in the UK today are heart disease, strokes and cancers. For men, the main cause of death is coronary heart disease, for women aged 35–49 it is breast cancer and for women over 80 it is dementia.

A. R. Omran's model of epidemiological transition (1971) outlines the complex long-term change in health and disease patterns, in which degenerative and man-made diseases gradually replace epidemics of infection as the main causes of morbidity and death. This model originally included three phases:

1 The Age of Pestilence and Famine

2 The Age of Receding Pandemics

3 The Age of Degenerative and Man-made Diseases

An additional fourth phase is recognised as The Age of Delayed Degenerative Diseases (delayed by modern technological and medical improvements). These shifts in health and disease patterns are closely associated with the demographic and socioeconomic transitions that accompany development.

Reasons for global contrasts in prevalence of communicable and non-communicable diseases

In LIDCs, communicable diseases (diseases of poverty) cause a higher proportion of deaths and morbidity than non-communicable diseases. Reasons for failure to eradicate vaccine-preventable diseases in LIDCs include:

- poverty of state governments — insufficient sustainable finance and lack of resources to increase the scale of interventions can cause epidemics, especially EIDs
- inadequate sanitation
- lack of access to clean drinking water
- limited access to education
- poor diets lead to malnutrition and **undernutrition** — deficiency of nutrients in the diet may cause rickets (vitamin D), beriberi (vitamin B1) and scurvy (vitamin C), and these diseases increase susceptibility to infections
- tropical or subtropical climates ideal for disease vectors, and an inability to remove stagnant water sources
- household poverty, poor quality housing and overcrowding

Reasons for the higher proportion of deaths and morbidity from non-communicable diseases (diseases of affluence) in ACs include:

- successful reduction or elimination of communicable diseases — this is partly the result of government wealth and affluence of households, enabling investment in medical advances and treatments
- high standards of living including sanitation, clean water supplies and good nutrition
- education and awareness of potential medical conditions

The prevalence of CVDs, cancers, chronic respiratory diseases and diabetes in ACs is linked to behavioural risk factors such as tobacco use, physical inactivity, harmful use of alcohol and unhealthy diet. In addition, air pollution and **overnutrition** contribute to these conditions. Increasing longevity leads to a higher percentage of deaths from degenerative diseases.

It should be understood that non-communicable diseases are an increasing problem in LIDCs. Globally, they account for a high proportion of deaths in low- and middle-income countries, where mortality from NCDs occurs increasingly at earlier ages.

Knowledge check 8

What is meant by epidemiological transition?

Undernutrition Too little food intake to maintain body weight.

Overnutrition Prolonged excessive food intake that increases body weight.

Exam tip

The relationship between disease and level of economic development can be illustrated and explained by the epidemiological transition of different countries.

This part of the specification requires a case study of any **one country** to illustrate the impact of air pollution on incidence of cancers. Possible choice of country could include India or the UK. For your chosen country you should be able to illustrate:

- causes of air pollution
- the impact of air pollution on incidence of cancers
- national solutions
- global solutions

Explain the difference between 'communicable' and 'non-communicable' diseases.

How effectively are communicable and non-communicable diseases dealt with?

The causes, impacts and success of mitigation and response strategies

This part of the specification requires **two** case studies to examine the causes of communicable and non-communicable diseases, their impacts and mitigation strategies. These should include a communicable disease in an LIDC or EDC (e.g. malaria in Ethiopia, cholera in Haiti) and a non-communicable disease in an AC or EDC (e.g. cancer in the UK or India). Each case study should illustrate:

- environmental and human causes of the disease
- prevalence, incidence and patterns of the disease
- socioeconomic impacts of the disease
- government and international agencies' direct and indirect strategies to mitigate against the disease and respond to outbreaks
- varying levels of success of these strategies

Prepare for exam questions that ask you to assess the effectiveness of different strategies used to mitigate the effects of particular diseases.

How far can diseases be predicted and mitigated against?

Increasing global mobility affects disease diffusion and response capability

Current action regarding global health can help to change the future disease burden. For example, reducing hunger, or improving nutrition, in childhood can help to mitigate the risks of communicable disease in adulthood.

The role of international organisations in combating disease

Strategies to predict and mitigate disease are central to the work of international organisations such as the WHO, UNICEF, UNAIDS, AMREF (African Medical and Research Foundation), international NGOs such as Médecins Sans Frontières, and CSOs such as the Bill & Melinda Gates Foundation. Increasing global mobility has the contradictory effects of contributing to diffusion of communicable diseases, but also enabling international organisations such as the WHO, which coordinates international strategies, to organise more rapid responses to disease outbreaks.

UNICEF is a leading international agency established in 1946 as a relief organisation for children. Among its aims today, it helps children (and their mothers) in developing countries whose lives are at risk. Its strategies include:

- cooperating with the WHO, state governments, CSOs and local communities to operate the Integrated Management of Childhood Illnesses (IMCI) strategy
- delivering Sustainable Development Goals (SDGs — including SDG 3, good health and well-being) and Millennium Development Goals (MDGs — especially MDG 4, reducing infant mortality, and MDG 6, malaria control)
- conducting research in assessment of shifting disease burdens — UNICEF monitors global and national data concerning child health
- combating HIV/AIDS through education programmes, helping prevent mother-to-child transmission
- providing financial and technical support
- intervention in healthcare and nutrition through immunisation, prevention and control of malaria, control and treatment of diarrhoeal and respiratory diseases, eradication of Guinea worm disease, preventing anaemia and promoting breast feeding

UNICEF has a major role in rapid response in emergencies, including:

- provision of fresh water and shelter
- supply of medicines, medical equipment, and vitamin A and micronutrients that help prevent malnutrition
- supply of insect-treated nets in malaria-infested areas

Further details are available at www.unicef.org.

A disease outbreak at a global scale

Severe acute respiratory syndrome (SARS) is a viral respiratory illness of zoonotic origin caused by the SARS coronavirus. It spreads by close person-to-person contact and through the air via respiratory droplets when an infected person coughs or sneezes.

Global spread of the SARS pandemic in 2002–3 involved 37 countries and led to 8,096 reported cases, including 774 deaths, many in Hong Kong. The first case was in Guangdong province in southern China in November 2002. Infected people spread the disease via international air travel, first to Vietnam, Taiwan and Singapore. Later, it spread to countries as diverse as Australia, Canada, Italy, the UK and the USA. There was further diffusion of the disease within China. The WHO declared it contained by July 9 2003 (see Figure 4). Details by country of the exact dates of first cases, numbers of cases and deaths can be found on the WHO SARS web page: www.who.int/csr/sars/country/en/.

UNICEF United Nations International Children's Emergency Fund.

> **Exam tip**
> You should be prepared to evaluate the contribution and roles of international organisations in the mitigation of disease, including the WHO, UNICEF and NGOs.

Pandemic An epidemic which spreads worldwide.

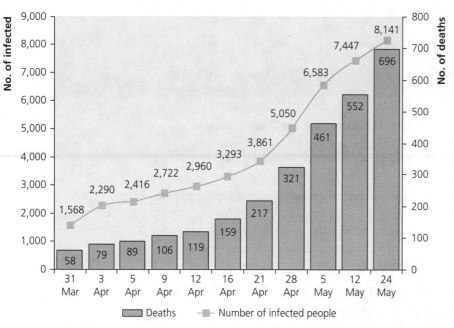

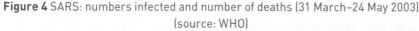

Figure 4 SARS: numbers infected and number of deaths (31 March–24 May 2003)
(source: WHO)

> **Exam tip**
>
> You should be able to analyse and evaluate quantitative and qualitative data, which may be presented in various different formats.

Case study

This part of the specification requires a case study of the role of **one NGO** in dealing with a disease outbreak in **one country**. Possible choices could include the British Red Cross working in Haiti following the 2010 earthquake and cholera outbreak. Information is available at www.redcross.org.uk. For your chosen example you must illustrate:

- causes and effects of the outbreak
- the NGO's response — this should include its overall strategy for the country and its specific work with local communities and households

> **Exam tip**
>
> Case studies are important to support points made in the more extended responses. Learn some key statistics and place-specific details to give your answers greater authority.

Mitigation strategies to combat global pandemics

Many barriers stand in the way, but governments and international agencies are working together to combat global pandemics.

The effects of physical barriers on disease mitigation strategies

Physical barriers affecting disease mitigation strategies include mountain ranges, large water bodies, large areas of aridity and areas of extreme climate.

Positive effects of physical barriers on mitigation include:

- limiting the spread of disease, since population movement is restricted
- restricting disease vector habitats
- reduced risk of infection due to lack of outside contact for isolated villages in rural peripheries, in mountain valleys or within tropical rainforest

- disease outbreaks occurring within an isolated community, which are more likely to be contained

Negative effects of physical barriers on mitigation include:

- restricting movement of medical assistance and emergency aid in the event of an outbreak
- delay of humanitarian response to disease outbreaks following natural hazards — for example, an earthquake may cause landslides, blocking roads or airports
- excess water as a result of river flooding or monsoonal rain, which may leave standing water that provides breeding habitats for mosquitoes and other water-borne, disease-carrying vectors
- flood waters restrict movement of aid workers, medicines, food and fresh water into severely affected areas
- lack of outside contact for isolated communities, which may reduce immunity — on contact with infected outsiders any introduced disease can spread quickly, affecting a high proportion of the population

Mitigation strategies of government and international agencies to combat global pandemics

The WHO states that:

> A pandemic is a worldwide epidemic caused by a novel virus that affects most or all age groups within a period of months.

Pandemic influenza emerges as a result of a completely new flu virus, which spreads more easily and with more serious effects, since global populations may be less immune to it.

The WHO attempts to mitigate pandemic influenza through its Global Influenza Programme (GIP). This provides member states with:

- assessment of risk, to devise responses and recovery actions based on surveillance and monitoring of previous outbreaks
- encouragement and advice for every country to create a national pandemic preparedness plan
- advice on selection of strains for new vaccine production under the Global Action Plan for Influenza Vaccines
- guidance on treatment of patients, establishing education programmes and use of rapid diagnostic tests for screening at a country's entry or exit points
- plans for high-density populations such as in refugee camps, including public health advice regarding social distancing, hand hygiene and household ventilation — this helps to delay the spread of influenza and reduces the number of deaths

The UK government preparedness strategy for pandemic influenza includes:

- public guidance for international travel
- guidance for business and work places
- coordination of collective action of government, essential services, businesses, media, other public, private and voluntary organisations and local communities
- encouragement of the public to follow government advice by adopting basic hygiene measures and reducing personal risk of catching or spreading the virus
- prediction of impacts such as the probable number of infected people, number of additional deaths, and social and economic disruption

Can diseases ever be fully eradicated?

Nature provides medicines to treat disease

Medicines from nature have been used to treat diseases for millennia. Approximately half of all human pharmaceuticals in use today are derived from natural sources, including plants, animals and microbes.

Medicines from nature: their habitats and growing conditions

Some compounds obtained from plants have been very significant in their effects, such as morphine, quinine (see Table 4), aspirin and cournadin. Other examples derived from nature include use of snake venoms in drugs to treat heart problems.

Table 4 Examples of medical drugs derived from natural compounds

Drug	Source	Growing conditions	Medical usage
Morphine	Dried latex from seed pods of several species of opium poppy	Warm, humid conditions. Clear sunny days with temperatures 30–38°C. Susceptible to frost and wet weather. Deep clay loam, well-drained soils rich in humus. Soil pH 6–8.	Pain reliever
Quinine	Dried bark of cinchonas evergreen tree	Average temperatures above 20°C. Humid conditions with annual rainfall in excess of 2,000 mm over at least 8 months. No frost. Well-drained fertile soils with abundant organic matter and good moisture-holding capacity.	Malaria — kills malarial parasites in red blood cells

Adapted from Table 11.6, p.359, Raw, M. et al (2016) *OCR A-level Geography* (2nd edn), Hodder Education.

> ### Case study
>
> This part of the specification requires a case study of **one medical plant**. Possible examples include rosy periwinkle, white willow tree, cinchona tree, opium poppy and foxglove. For your chosen example, you should be able to illustrate:
> - growing conditions
> - medicinal importance for treatment of disease
> - international trade
> - sustainability of its use

Conservation issues relating to international trade in medicinal plants

The number of species and quantities of plant material in medicinal use are significantly large. According to Botanic Gardens Conservation International, over 400 medical plants are at risk of extinction. Examples include yew trees (bark is widely used in producing cancer drugs) and magnolia (traditional Chinese medicine used to fight cancer, dementia and onset of heart disease).

There is growing demand for plants for herbal and pharmaceutical medicines from tropical and subtropical countries in Africa and southeast Asia. For example, the bark of *prunus Africana*, the African cherry, is valuable in the treatment of benign prostatic hypertrophy. Demand for the bark to produce medical drugs is high in Europe, but supply from countries such as Cameroon is environmentally and socially unsustainable.

Conservation issues arising from international trade in medicinal plants include:
- over-harvesting of wild plants rather than cultivated sources for herbal medicines and manufacture of pharmaceutical drugs
- obtaining samples for continued research to develop new medical products
- habitat loss, often caused by deforestation of tropical rainforest
- diminishing biodiversity
- endangering species survival
- disruption or loss of natural ecosystems

Treaties such as the Convention on International Trade of Endangered Species of Wild Flora and Fauna address conservation issues. Large pharmaceutical transnational corporations (TNCs) have also adopted increasingly responsible policies.

Top-down and bottom-up strategies that deal with disease risk and eradication

Both top-down and bottom-up (grassroots) strategies are being employed to deal with disease risk.

Strategies for disease eradication at global and national scales

Eradication of a disease is the permanent reduction to zero of its incidence worldwide. The WHO declared smallpox eradicated in 1980. This was accomplished through surveillance, quickly identifying new smallpox cases and a vaccination programme in which anyone exposed to smallpox was found and treated to prevent further spread.

Guinea worm disease is transmitted to humans when people drink stagnant water contaminated with parasite-infected water fleas. According to the WHO, Guinea worm disease is eradicable because:
- diagnosis is relatively easy
- the host is restricted to stagnant water bodies
- intervention is simple, cost-effective and easy to implement
- geographical distribution is limited
- transmission is seasonal
- there is no animal reservoir
- LIDC governments are willing to provide political backing and financial commitment

Strategies of the WHO and partners include:
- mapping all endemic villages
- encouraging community-based surveillance for immediate detection and reporting
- intervention to ensure access to safe water, health education and vector control
- establishing global databases to monitor the epidemiological situation
- managing certification for global eradication country by country

The International Task Force for Disease Eradication (ITFDE) evaluates the potential for eradication of infectious diseases. Eight diseases have been identified: Guinea worm, polio, mumps, rubella, lymphatic filariasis, cysticercosis, measles and yaws.

Exam tip

You should be able to identify and explain conservation issues, including the sustainability of ecosystems, which have arisen as a result of increased international trade in medicinal plants.

Case study

This part of the specification requires a case study of the global impact of **one pharmaceutical TNC** (e.g. GlaxoSmithKline). For your chosen TNC you should be able to illustrate the company's:
- global structure and operations
- research and development (R&D), including any scientific breakthroughs
- patented drugs
- drug manufacture
- global flows of distribution

Elimination of a disease is the reduction to zero of its incidence in a defined geographical area. The ITFDE has identified river blindness, trachoma and schistosomiasis for elimination. In 2012, the WHO removed India from the list of polio-endemic countries. In 2015, maternal and neonatal tetanus were also eliminated in India — actually less than one case per 1,000 live births in every district across India. This was the result of the Indian government's commitment to immunisation, antenatal care services and provision of skilled birth attendance, even in populations vulnerable to poverty or isolation. Many women were involved locally, training as village auxiliary nurse midwives and birth attendants.

Impact of grassroots strategies in combating disease risk

Top-down strategies imposed by international organisations and national governments are public health interventions that rely on availability and sustained use of resources. This approach can lead to substantive change in disease risk for an entire area, such as a country or region. However, not all countries can afford to sustain this strategy in terms of staff needed, training and provision of medical facilities. It can also lead to dependence on government services.

Bottom-up strategies are considered to be more effective interventions since they are self-motivated among local communities and families, where need is greatest. This is a more sustainable approach, since local people have a vested interest in accepting responsibility for their own health. It is achieved at grassroots level through education and the assistance of trained members in local communities. In many LIDCs and EDCs, where gender inequalities restrict development, changing roles and empowerment of women have led to increased female involvement in family and community health.

Knowledge check 10

Explain how the terms 'eradication' and 'elimination' of a disease are used in epidemiology.

Exam tip

Remember that strategies used to minimise the impacts of disease operate at different scales from global to local.

Knowledge check 11

What are 'top-down' and 'bottom-up' strategies for disease control?

Summary

- The incidence of specific diseases can be mapped to show their spatial distribution at different scales from local to global.
- Physical, economic, social and political factors influence the spread of disease.
- Climate change can lead to re-emergence of infectious diseases and their diffusion.
- Natural hazards can lead to the outbreak and spread of diseases.
- There is a relationship between disease and level of economic development. Epidemiological transition involves a decrease in prevalence of communicable diseases and an increase in prevalence of non-communicable diseases. These changes are explained by interaction of physical, economic, social and political factors.
- International organisations and national governments implement a range of strategies to mitigate against both communicable and non-communicable diseases, with varying degrees of success.
- The WHO and other international organisations, including NGOs, have developed strategies for response to, and combating the spread of, disease outbreaks.
- Physical barriers of the natural environment have positive and negative impacts on strategies to mitigate against pandemics.
- Medicines derived from nature have been in use for millennia. Increased demand for medical plants and their international trade have threatened their sustainability.
- Strategies for disease eradication include 'top-down' and 'bottom-up' approaches.

Exploring oceans
What are oceans' main characteristics?

Oceans and seas make up about 71% of the Earth's surface.

The world's oceans are a distinctive feature of the Earth

There are five oceans and many seas. The largest, the Pacific (155.6 million km²), is just over twice the area of the Atlantic, in second place. The smallest ocean, the Arctic (14.1 million km²), is nearly five times the size of the largest sea, the Mediterranean (3.0 million km²).

Ocean basin relief

Oceans share the same basic structure (see Figure 5).

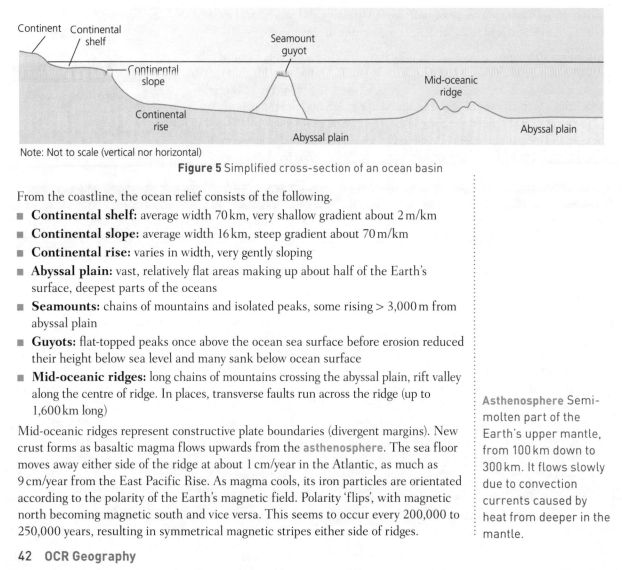

Note: Not to scale (vertical nor horizontal)

Figure 5 Simplified cross-section of an ocean basin

From the coastline, the ocean relief consists of the following.

- **Continental shelf:** average width 70 km, very shallow gradient about 2 m/km
- **Continental slope:** average width 16 km, steep gradient about 70 m/km
- **Continental rise:** varies in width, very gently sloping
- **Abyssal plain:** vast, relatively flat areas making up about half of the Earth's surface, deepest parts of the oceans
- **Seamounts:** chains of mountains and isolated peaks, some rising > 3,000 m from abyssal plain
- **Guyots:** flat-topped peaks once above the ocean sea surface before erosion reduced their height below sea level and many sank below ocean surface
- **Mid-oceanic ridges:** long chains of mountains crossing the abyssal plain, rift valley along the centre of ridge. In places, transverse faults run across the ridge (up to 1,600 km long)

Mid-oceanic ridges represent constructive plate boundaries (divergent margins). New crust forms as basaltic magma flows upwards from the asthenosphere. The sea floor moves away either side of the ridge at about 1 cm/year in the Atlantic, as much as 9 cm/year from the East Pacific Rise. As magma cools, its iron particles are orientated according to the polarity of the Earth's magnetic field. Polarity 'flips', with magnetic north becoming magnetic south and vice versa. This seems to occur every 200,000 to 250,000 years, resulting in symmetrical magnetic stripes either side of ridges.

Asthenosphere Semi-molten part of the Earth's upper mantle, from 100 km down to 300 km. It flows slowly due to convection currents caused by heat from deeper in the mantle.

Some ocean margins have different features, as **subduction** is taking place along them. A key feature is an ocean trench, the deepest being the Mariana Trench in the northwest Pacific, at between 7–11 km depth.

Horizontal and vertical variations in ocean water

Salinity

In general, **salinity** is higher around the tropics and lower towards the poles. Factors such as sea water evaporation and sea ice formation increase salinity. Salinity decreases with inputs of fresh water from major rivers, precipitation into the oceans and ice melt. It also changes with depth but is not a straightforward relationship. In some places such as the South Atlantic, higher salinity exists at the surface but by about 750 m has decreased significantly through the **halocline**. In other places, salinity increases through the halocline but once below about 1,500 m, it generally hardly changes with depth.

Variations in salinity affect water density, which in turn influences water movements. How, when and where water moves, both vertically and horizontally, are very important aspects of the oceans that affect ecosystems and human activities.

Temperature

Oceans act as vast reservoirs of heat energy. Much energy is required to lift water's temperature, but once heated, water retains that heat better than almost any other substance. A gradient in sea surface temperatures (SSTs) exists from the warmer tropics to the colder polar oceans. With increasing depth, water temperatures decline. Just below the surface is the **thermocline**, but below about 1 km water temperature hardly changes.

Warm and cold ocean surface currents

The overall pattern of surface currents is of warm water flowing north and south from the low latitudes to the middle and high latitudes. These flows transfer vast quantities of heat energy and are key influences on climates, ecosystems and human activities.

Some flows of water, known as gyres, are generated by winds. In the Atlantic and Pacific there are north and south gyres, while a southern one exists in the Indian Ocean.

The circulation in the North Atlantic

From the Gulf of Mexico, a relatively warm surface current, the Gulf Stream, moves northeastwards. In its most northern section it is known as the North Atlantic Drift. As it flows north of the Arctic Circle, its temperature and density fall so that the current sinks. This results in a deep, cold water flow back towards the low latitudes.

The cold Labrador Current extends from the Arctic Ocean southwards off the coasts of northeast Canada and the USA. A small branch of the Gulf Stream pushing northwards keeps the west coast of Greenland ice-free. An input of relatively warm and salty water flows out from the Mediterranean through the Straits of Gibraltar at a depth of about 1,000 m. Eventually, its density equals that of the surrounding water and mixing takes place.

Subduction Occurs where the oceanic crust sinks below either a continental or oceanic plate along a destructive boundary (convergent margin).

Salinity Measures the concentration of sodium chloride (salt) expressed as grams (parts) of sodium chloride per 1,000 g of water. Fresh water is < 0.5 parts per thousand (ppt), sea water average is about 35 ppt.

Halocline The sharp change in salinity with depth from just below the surface down to about 750 m. The rate of salinity change and vertical dimensions vary from place to place within oceans.

Thermocline The sharp change in temperature with depth from just below the surface down to about 1,000 m.

Knowledge check 12

What is the importance of the halocline and thermocline to the vertical movement of water in the oceans?

> ### Case study
>
> For this part of the specification you are required to have a case study of **one renewable biological resource** within oceans, such as krill or whale. Your case study must illustrate:
> - the use and management of the resource
> - how the stakeholders' values, attitudes, socioeconomic status and political context influence the resource's use and management
> - the resource's resilience and key thresholds to initiate management

The use of ocean energy and mineral resources is a contested issue

Humans have relatively recently begun to exploit ocean energy and minerals. With advances in technologies and growing demand, pressure on such resources has been increasing.

Non-renewable ocean resources: oil and gas

Oil and (natural) gas are finite and **non-renewable**. Rising demand for oil and gas energy from the mid-twentieth century gave an impetus to extend exploration and production into deeper and stormier sea areas. Commercial drilling rigs operate in depths up to 2,000 m+. Locations such as the Gulf of Mexico, the Persian Gulf and the North Sea produce large quantities of **hydrocarbons**. The exploitation of these energy resources is sensitive to their price. Because oil and gas prices vary considerably through time, investment in marine energy industries can be volatile.

There is considerable onshore investment related to the exploitation of marine oil and gas. Docks, ship and rig building yards, terminals where energy comes ashore, storage facilities, pipelines and refineries are all part of the marine energy industry. The increasing scale of vessels used to transport oil and gas require particular features of a coastal area. A supertanker of approximately 200,000 tonnes needs not less than 30 m depth of water and 2 km+ of open water in which to turn.

The exploitation of marine oil and gas is a contested issue.

Positive impacts include:
- employment opportunities
- wealth creation
- raw materials for a wide range of products, e.g. fertilisers, paints, plastics, medicines
- rigs that act as artificial reefs, increasing local marine species populations

Negative impacts of oil and gas exploitation include:
- over-dependence of local communities on this one industry
- visual impact, e.g. storage tanks and refineries
- oil spills
- ecosystem disturbance, e.g. underwater noise

Non-renewable resources Once used, these resources cannot be replaced in human timescales.

Hydrocarbons The main chemical compounds making up fossil fuels.

> ### Exam tip
>
> Knowledge and understanding of the management of ocean energy resources is helped through a case study, e.g. the Gulf of Mexico or North Sea.

Renewable ocean resources: tides and waves

Alternative marine renewable energy sources are being investigated to reduce the 75–80% of global energy consumption coming from fossil fuels.

Tidal energy

Tidal and wave energy are **flow resources**. Tidal energy comes from the rise and fall of water as the tidal cycle operates. Mills powered by tides existed for centuries until fossil fuels became widespread from the late eighteenth century. The potential of tidal power depends on local physical geography such as the shape of the coastline and **tidal range**.

Tidal power is contested, as both advantages and disadvantages exist. Tides are reliable, regular and predictable. Most coastlines experience two high and two low tides approximately every 24 hours. However, a significant disadvantage is that few suitable locations are close to places where electricity demand is high enough to justify the huge cost of developing a tidal power station.

One way to exploit tides is via a **barrage**. There are several tidal schemes in operation, the largest being the Shiwa Scheme in South Korea. Around the UK, a number of schemes have been proposed but none yet built.

Wave energy

The rise and fall of water as a wave passes has great potential for energy generation, much more so than tidal. However, developing a practical, reliable system has proved a significant challenge.

Many systems are under development and in all examples wave energy creates hydraulic pressures, which drive electricity generators. So powerful is wave energy at the locations that offer regular, large waves that the equipment has to be very large in order to be robust. So far the results mean that too little of the wave energy is converted to actual usable energy.

Economically, tidal and wave energy are not viable at present but over the course of the next few decades, depending on the cost of energy and political attitudes, such schemes are likely to receive increasing attention.

Sea-floor mining: the final frontier?

Underwater mining for minerals has received attention, as scientists have discovered both ferrous and non-ferrous mineral deposits. Additionally, diamonds are mined off the coast of southern Africa.

Forty years ago there was some recovery of manganese nodules from the sea bed. However, it was an expensive operation and therefore not commercially viable. But increasing commodity prices in the late twentieth and early twenty-first centuries have reignited interest in sea-floor mining. The growing importance of **rare earth elements (REEs)** to technologies such as mobile phones has boosted research into sea-floor mining. Because REEs now play a pivotal role in much of the technology humans rely on, their supply is a geopolitical issue. China dominates their production, which is a concern for nations such as the USA, Japan and in Europe.

Flow resources Renewable natural resources that can be used and replenished at the same time.

Tidal range The vertical difference in height between consecutive high and low waters over a tidal cycle.

Barrage A dam built across a bay or estuary. Water is allowed through pipes in the dam as the tide rises and falls, turning turbines as it flows and generating electricity.

Exam tip

Knowledge and understanding of the use of tidal power is helped through a case study, e.g. the Swansea Bay or Severn Estuary schemes.

Rare earth elements (REEs) Metals whose properties are vital to miniaturisation in electronic products, telecommunications and military hardware.

Deep-sea remotely operated vehicles (ROVs) are deployed to survey and sample sea-floor deposits in parts of the Pacific Ocean. The geopolitical concerns might influence whether permission is given or not to prospect on the sea floor. This is another contested issue, as there are concentrations of minerals around mid-ocean ridges and hydrothermal vents. We have surveyed so little of the deep ocean that mining may cause serious damage to unique ecosystems. However, some sea-floor minerals exist at much higher concentrations than occur on land, so less volume of material is required — land-mined copper ore is typically about 4% metal, whereas sea-floor ore can be ten times more concentrated.

Mining creates **tailings**, which are released back into the ocean. This causes cloudy water and when the sediment settles, it can smother sea bed ecosystems such as corals.

Governing the oceans poses issues for the management of resources

The concept of the 'tragedy of the commons' applies to the oceans. When a resource is seen as belonging to all, such as the oceans or atmosphere, there is a tension between the interests of everyone, the common good and self-interest. People tend to exploit the resource without considering their impact on it. The advantage to the individual is greater than its cost, as the cost is shared among very many.

How is the ocean owned?

Various frameworks for managing oceans have been developed. The key one is the United Nations Convention on the Law of the Sea (UNCLOS), which attempts to define marine rights and responsibilities. It identifies various zones with decreasing national rights as distance from the land increases (see Figure 7).

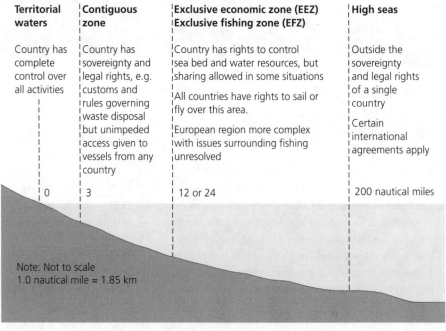

Territorial waters	Contiguous zone	Exclusive economic zone (EEZ) Exclusive fishing zone (EFZ)	High seas
Country has complete control over all activities	Country has sovereignty and legal rights, e.g. customs and rules governing waste disposal but unimpeded access given to vessels from any country	Country has rights to control sea bed and water resources, but sharing allowed in some situations All countries have rights to sail or fly over this area. European region more complex with issues surrounding fishing unresolved	Outside the sovereignty and legal rights of a single country Certain international agreements apply
0	3	12 or 24	200 nautical miles

Note: Not to scale
1.0 nautical mile = 1.85 km

Figure 7 The UNCLOS defined zones for ocean management

Tailings Mining residue produced when rock is washed to separate the metal ore from waste material.

Knowledge check 14

Why is the exploitation of so many ocean resources a contested issue?

Most countries have signed the convention but issues persist in some locations, as neighbouring nations dispute who owns what. This is particularly sensitive when resources such as fish or minerals are involved. As such, UNCLOS doesn't cover several management issues, such as fishing in the deep oceans, regulation of underwater noise or patenting genetic material through **bioprospecting**.

The International Seabed Authority is intended to oversee activities such as sea-floor mining. Hydrothermal vents, however, are not included in the authority's responsibilities. The International Whaling Commission is intended to supervise the management of whaling. Its 88 member countries, however, do not always agree and although a moratorium banned commercial whaling, it persists and is still a highly contested matter.

Marine reserves

Marine reserves are the ocean equivalent of national parks on land. They are established at locations where unique biological, geological, historic or cultural features exist. They protect areas for the future and are seen as significant to developing the oceans' resilience to impacts from climate change, such as water warming and ocean acidification. Currently some 3% of the world's oceans are designated as marine reserves. The International Convention on Biological Diversity is aiming for 10% by 2020.

How and in what ways do human activities pollute oceans?

Pollutants that affect the ocean system

Point-source **pollution** comes from a clearly identifiable location, such as an oil rig. Non point-source pollution occurs from several dispersed locations, such as ship exhaust fumes. Pollution of oceans also originates from the land — winds and rivers carry substances such as fertiliser and **particulates** to oceans. It is now recognised that pollution badly affects the oceans, despite their vast volume. Some pollutants are so toxic and persistent that their effects are felt for decades.

Fossil fuel burning at sea

Shipping relies on burning fossil fuels, which produces pollutants such as carbon dioxide, sulphur dioxide and particulates. Because there are 90,000–100,000 ocean-going cargo vessels and distances sailed are long, air pollution from ships is globally significant. Most ships burn cheap, low-quality fuel, which is particularly high in sulphur. The largest vessels emit about 5,000 tonnes of sulphur a year. Annually, sea transport accounts for 9% of sulphur dioxide and 15–30% of nitrous oxide pollution globally. Annual estimates suggest that 3.5–4% of all human-activated GHGs are from shipping.

When ships are in coastal waters and ports, their pollution directly threatens human health. Air pollution is two to three times higher in the areas close to large ports, such as Long Beach, California. The growth in cruise-liner traffic has raised levels of air pollution in locations such as the steep-sided fjords of Norway and Alaska, and ports in the Caribbean.

Bioprospecting
The discovery and commercial exploitation of biological material, e.g. genetic material in organisms.

Exam tip
Knowledge and understanding of marine reserves is helped through a case study, e.g. St Kilda (Western Isles) or Chagos (Indian Ocean).

Pollution When a substance is added to the environment that affects organisms adversely and at a rate greater than the environment can cope with.

Particulates Very small particles such as dust, soot and salt crystals. Some occur naturally (e.g. from volcanoes) but others originate from human activities (e.g. fossil fuel burning).

Regulation of ship emissions is relatively easy but enforcement is difficult. On-ship monitors are being developed as is increasing the fuel efficiency of marine engines using fuel with lower sulphur content. Improved ship design, such as more efficient hulls and propellers, can save fuel. Some shipping companies have been experimenting with cutting the speed of their vessels to reduce fuel costs but they need to balance this with increased costs from longer journey times.

Domestic and industrial pollutants

Rivers discharge dissolved chemicals or solids such as heavy metals and plastics into the sea. Pollutants released into the atmosphere can be carried out over the sea, where they fall via rain.

Over the centuries, people have viewed the sea as a convenient dumping ground for vast amounts of pollution. Most ACs have now significantly reduced the practice of discharging raw sewage into coastal waters. Incidents still do occur, but generally coastal water quality has improved. Likewise, industrial effluents are strictly regulated in ACs to prevent pollution. Progress along these lines is observable in EDCs but environmental concerns do not yet have a high priority. In most LIDCs, large volumes of pollutants are released into rivers and the sea.

Algal blooms are a sign that water quality has deteriorated. The addition of large amounts of nitrates and phosphates, from fertilisers or detergents for example, create the conditions in which algae can flourish. The level of dissolved oxygen reduces and can result in high levels of mortality in aquatic organisms. Human health is also affected, as well as regional and local economies, for example the closure of shellfish industry until the ecosystem recovers.

> **Algal blooms** Occur when colonies of algae (simple plants) grow out of control. They occur naturally but are closely associated with human activities.

Nuclear waste

For much of the second half of the twentieth century, oceans were seen as suitable locations to dispose of nuclear waste. It was assumed that radioactivity would disperse to low levels in the vast volumes of water. In addition to deliberate acts, unintentional release has occurred. A handful of nuclear-powered submarines are lying on the sea bed, slowly decaying. Eventually, the shields surrounding their reactor cores will disintegrate, releasing high-level radioactive material.

In 2011, the Fukushima nuclear power plant in Japan was overwhelmed by a tsunami, with the result that substantial quantities of radioactivity were leaked into the northwest Pacific. Marine food chains have been contaminated, resulting in bans on shellfish and fish harvesting in the local area.

> **Knowledge check 15**
>
> Why is radioactive waste in oceans such a concern?

Offshore oil production and transport pose threats to people and the environment

Many of the planet's most diverse and ecologically important regions hold large underground deposits of oil and gas. Extracting these deposits has resulted in damage to the environment and people.

> **Case study**
>
> For this part of the specification you are required to have a case study of **one oil spill**, such as from a tanker running aground or from a drilling rig. Your case study must illustrate:
> - impacts on the physical environment and marine ecosystems
> - impacts on human activities, such as fishing and tourism
> - management of the oil spill and its impacts

The pattern of ocean currents can disperse and concentrate pollution

All sorts of debris enters the marine environment from rivers, beaches and vessels, either accidently or deliberately. Some of the debris biodegrades relatively quickly, but increasing quantities are persistent, potentially over many decades. Marine debris poses many dangers. Large creatures, such as sharks and seals, can become entangled in discarded nets and drown. A container washed off a ship will float and is a serious hazard for smaller vessels. Perhaps the most widespread issue relates to plastics. Most do not biodegrade but break into very small pieces called microplastics, which organisms can ingest. The plastic passes through food chains and webs, accumulating in higher trophic levels. This causes health issues, such as preventing digestion, and some of the chemicals are toxic when accumulation reaches high levels.

> **Case study**
>
> For this part of the specification you are required to have **one** case study of the accumulation of plastic in **one ocean gyre**, such as the Northern Pacific gyre, sometimes called the 'Great Pacific Garbage Patch', or the Indian Ocean including the Bay of Bengal. Your case study must illustrate:
> - the causes of the accumulation of plastic in the ocean gyre
> - impacts on marine ecosystems

Exam tip
You must have secure knowledge and understanding of the pattern of ocean currents in order to understand the dispersal and concentration of pollutants.

Ocean gyre A system of circular currents formed by wind patterns and the planet's rotation.

How is climate change impacting the ocean system?

There is overwhelming evidence that the Earth's climate is currently undergoing significant change and warming.

Climate change is altering the nature of the ocean's water

Because the ocean and atmosphere systems are strongly interlinked, oceans are also changing. This occurs through several processes.

Ocean acidification

Oceans are important **sinks** within the Earth's carbon cycle. They have absorbed some 30% of the anthropogenic carbon dioxide generated over the past 250 years,

Sink Anything that accumulates more of a particular substance than it releases, e.g. the oceans act as a sink for carbon dioxide.

which affected their chemistry. Most significant is the rapid change in their pH level — average global ocean surface pH has fallen, resulting in a 30% rise in the acidity of ocean water since about the mid-eighteenth century. By 2100, the forecast is that acidity will have doubled.

Impacts on marine ecosystems

Many marine organisms build skeletons that require them to accumulate calcium carbonate. Increasing ocean acidity reduces this ability, and organisms such as zooplankton are less likely to reach maturity and breed, and are more susceptible to predation. Given that many of these organisms make up lower trophic levels, a collapse in their numbers will feed back throughout the oceans. Many species of fish rely on zooplankton and if fish numbers fall, so will those organisms that prey on these fish. The same risk applies to shellfish such as mussels and oysters.

There is also an impact on the carbon cycle. When marine organisms die their skeletons fall to the sea bed, accumulate and eventually become sediment. This stores carbon for millions of years.

Some species may thrive in more acidic waters. Research has suggested that some jellyfish may increase significantly in numbers and disturb the balance of an ecosystem.

Zooplankton Tiny organisms living in the top 200 m of the oceans.

Impacts on people

Ocean acidification threatens harvesting of both wild and farmed fish as well as shellfish. This will put the food security of millions at risk. About 200 million tonnes of seafood are produced annually, just under half from the Pacific and about one-quarter from the Atlantic.

Significantly, some countries most dependent on seafood for protein include LIDCs and EDCs — in places such as The Gambia, Bangladesh and Pacific islands, seafood can make up 50% of dietary protein. There are limited agricultural opportunities for many people, especially in the context of rising numbers of people and shortages of land.

Many ACs also gain provisioning services (see Table 5 on p. 45) from marine harvests, some substantially so, such as Canada, Iceland and Japan.

Warming oceans and coral ecosystems

Corals tend to live in symbiosis with types of algae called zooxanthellae, which give coral its colours.

The algae release nutrients via photosynthesis, which polyps feed on. In return, algae are protected by the coral's skeleton and obtain some minerals from this. As oceans warm, stress increases on both coral and algae. When this stress crosses a critical threshold, the algae are expelled and coral bleaching occurs. At some locations, up to 80% of the coral has been affected, for example the Maldives and parts of the eastern Pacific. Some reefs have shown signs of recovery and research, including the use of satellite technologies, continues into this phenomenon. One aim is to understand in what ways humans can aid the resilience of coral to stress from warming water.

Corals Invertebrates called marine polyps, a few millimetres in diameter and a few centimetres in length. They secrete a protective calcium carbonate skeleton around themselves.

Symbiosis The close association between two or more organisms of different species that tends to benefit each member.

Threats to biodiversity

Reefs are known as the 'rainforests of the oceans' as they are home to so many species — more than 25% of all marine life. Climate change threatens reefs due to:

- coral bleaching
- rising sea level, which increases water depth, reducing light levels
- increased wave energy from more intense storms, which mechanically damages reefs
- ocean acidification, which reduces coral's ability to build carbonate structures

Threats to local communities

There are direct and indirect threats to local communities. Reefs:

- absorb wave energy thus protecting land behind them
- are where fish breed and develop, which are then protein sources
- offer tourism opportunities, which support local communities

Climate change is altering sea levels

Sea level is not stationary, particularly over geological time. In the past 2 million years, sea level has varied greatly, once to 120 m below its present level.

Two types of change affect sea level: eustatic (a worldwide absolute change in sea level, as all oceans and seas are connected) and isostatic (absolute changes in the level of the land — these are often localised and the result of tectonic forces).

Recent sea-level change

The effects of global warming on sea level are a key area of research for the **IPCC**. It is now clear that a eustatic rise in sea level is underway, averaging about 3.0 mm/year at the start of this century. Advances in technologies such as satellite observations allow increasingly accurate and reliable measurements.

The causes of recent sea-level change include the following:

- **Thermal expansion of water:** as water temperature rises, its density decreases and its volume increases. To date, this is the primary cause of sea-level rise.
- **Melting of glaciers/ice caps, e.g. in the Andes and Himalayas:** as the atmosphere warms, ice melts and the water released flows into the oceans.
- **Melting of Antarctic and Greenland ice sheets:** a complex picture made up of the margins melting and thinning while the central areas are slightly thickening. Meltwater flows into the oceans.

> **Exam tip**
>
> Although a case study of the threat to coral from climate change is not directed in the specification, researching an example will give you knowledge and understanding of what changes are actually taking place and the management measures being introduced. You can find excellent material online for locations such as the Great Barrier Reef, Australia, or reefs in the Indian Ocean or the Caribbean.

The **IPCC** is a large international and interdisciplinary group of scientists researching the causes and effects of climate change on the environment and human societies.

> **Case study**
>
> For this part of the specification you are required to have a case study of **one island community** in either the Indian (e.g. Maldives) or Pacific oceans (e.g. Fiji) or the Caribbean Sea (e.g. St Vincent and the Grenadines). Your case study must illustrate the:
> - threats to island communities
> - impacts on communities
> - adaptations by governments and communities in both the short and long term

Climate change is altering high-latitude oceans

Climate change in high-latitude oceans is bringing about dramatic changes in their physical characteristics, which could cause irreversible damage both to their ecology and to the Earth as a whole.

Why does sea water freeze at high latitudes?

Due to the salt it contains, sea water freezes at −2.0°C. When it freezes, the ice contains virtually no salt, as it is the water that freezes.

In the Arctic and Antarctic Circles, surface heating is less intense than in mid- and low latitudes. The sun's rays strike the surface at low angles, meaning that for the same amount of incoming energy, the area heated is much greater in the high latitudes. Snow and ice surfaces have high albedo, further reducing temperatures.

Albedo The proportion of sunlight reflected from a surface.

In the Arctic, winter temperatures average between −30°C and −35°C, while coastal regions of Antarctica are between −10°C and −30°C in its winter and close to freezing during summer months.

The relationship between global warming and sea ice

Two key aspects of sea ice are relevant: area and thickness. There is a clear trend of reducing sea-ice cover, and we are seeing record lows in the twenty-first century.

Military submarines operating under the Arctic sea ice provide data that show a steady thinning. In the 1960s, ice was typically 4 m thick whereas today it is about 1.25 m. The occasional years when sea ice increases do not negate the evidence of a sustained decline in ice area and thickness. New 'normals' are being established, which might lead to ice-free summers in the future. Perhaps most serious is that a point will be reached when a threshold is crossed. After that, so much of the Arctic Ocean is open sea that the water will absorb increasing amounts of solar radiation, leading to further ocean warming. In turn, more sea ice will melt and so an irreversible cycle will exist.

Knowledge check 16

Why is the Arctic region becoming an area of rising geopolitical tension?

Case study

For this part of the specification you are required to have **one** case study of the Arctic Ocean. Your case study must illustrate the:

- geopolitical implications of changes in ice cover in the region
- impact on indigenous people
- threats and opportunities posed by the opening up of ocean route-ways and increasing access to ocean-bed minerals
- management of the changing geography of the Arctic through international organisations

How have socioeconomic and political factors influenced the use of oceans?

Oceans are vital elements in the process of globalisation

For centuries, ocean transport has carried people and goods around the globe. Globalisation over the past 50 years or so has become more intensive. Links are longer and increasingly extensive, with more people and goods crossing oceans. Total world trade now accounts for about 45% of global GDP with much of this using ocean transport. Billions of people are linked in a vast web of interdependence. A key aspect of globalisation is time-space compression (a set of processes leading to a 'shrinking world'). Technology, especially communications and transport, has played a leading role in this process.

Global shipping routes

Most cargo vessels follow a network of core and secondary routes. An east–west corridor links North America, Europe and Pacific Asia via the Suez Canal, the Straits of Malacca and the Panama Canal. A core route also extends from Europe to eastern South America.

Factors influencing global shipping routes

The factors influencing global shipping routes include:

- the shape of coastlines
- winds
- ocean currents
- water depth, including the presence of offshore reefs
- sea ice and icebergs

The Suez and Panama Canals influence routes, as they offer very significant shortcuts — about 9,000 km and 10 days in the case of the Suez Canal, and about 13,000 km and 20 days for the Panama Canal. Both canals are being upgraded to allow more and larger vessels through.

Physical geography strongly influences port location, with water depth, tidal range and degree of shelter all key factors. Natural harbours, such as Sydney and Singapore, have developed into global shipping hubs. Extensive engineering, such as docks, breakwaters and dredging, have allowed ports to grow in locations less naturally favoured, such as Europoort, an extension of Rotterdam at the mouth of the Rhine in the Netherlands.

The direction and type of trade across the oceans

The size of market served by a port strongly influences the volume of goods traded through it. Europoort serves much of western and central Europe. Total population gives some indication of market size, but the income of those people is also significant. Europe, North America and Japan have wealthy populations, and their demands and abilities to pay for goods are considerable. The pattern is dynamic, as the growing middle classes in India and China, both numerically large groups, are creating rising demand for goods.

Globalisation The growing integration and interdependence of people's lives in a complex process with economic, social (cultural), political and environmental components.

Time-space compression A set of processes leading to a 'shrinking world'. This is caused by reductions in the relative distance between places, e.g. the cutting of journey times for goods and people.

Primary goods such as crude oil, minerals and agricultural products are shipped around the globe in vast quantities. Manufactured goods also account for significant flows. It is not, however, a simple pattern of raw materials coming from LIDCs and some EDCs to ACs and manufactured goods going in the opposite direction — ACs and EDCs are the leading exporters and importers of agricultural products, fuels and mining products.

Marine technology: a revolution in transport

One significant contributory factor to globalisation has been advances in technologies associated with ocean trade. **Containerisation** of cargoes reduces costs at every stage from factory to customer. Ships now spend less time in port, as they can be efficiently loaded and unloaded. Vessel size has increased dramatically over the past 50 years with the largest container ships carrying up to 18,000 individual containers. The **economies of scale** created by these changes also promote globalisation.

Bulk goods, such as oil, mineral ores and cereals, are transported in enormous vessels. The largest oil tankers carry about 3 million barrels of oil, equivalent to some 440,000 tonnes. Improvements in dock facilities have proceeded hand in hand with increases in ship size, with handling of all types of goods highly mechanised. One of the world's leading ports, Singapore, handles some 33 million containers a year.

Cruise liners have increased in number and size. The largest carry 6,000 tourists and 2,300 crew on a vessel of 225,000 tonnes and 360m in length. Many cruises consist of inhabitants of ACs visiting LIDCs or EDCs, such as in the Caribbean. Some AC locations receive significant numbers of cruise ships, such as the fjords of Alaska and Norway.

Submarine cables: unseen connections

Submarine cables were first laid across the Atlantic in 1866. By the early twentieth century, a global telegraph network of underwater cables had been installed. Telephone cables were laid in the 1950s and today the internet could not function without the network of fibre-optic cables that criss-cross oceans.

Oceans are important spaces where countries challenge each other

Oceans have long been the setting for rival countries to challenge each other. Geopolitics is especially contentious at sea as it is difficult to establish clearly defined boundaries. The establishment through UNCLOS of areas such as 'exclusive economic zones' continues to be contested. The Arctic is one region where tensions are likely to rise in line with the effects of global warming.

Naval power continues to be significant to **superpowers** such as the USA, and challenger states, such as China and India, see naval power as important to their growing role on the global stage.

China's growing naval power

For centuries, China has been a naval power within Asia. As China develops its regional power within Asia and the Pacific and extends this globally, it is investing

> **Exam tip**
>
> Research trading patterns so that you can quote figures and countries and/or trading blocs involved.

Containerisation The transport of goods in standard-sized metal boxes, allowing mechanised handling of large volumes of goods as well as increasing their security and safety.

Economies of scale In ocean shipping, savings in unit costs (i.e. per container) that arise from large-scale handling and transport of goods — doubling the number of containers does not double the cost of moving them, e.g. the crew doesn't need to be twice as numerous.

A **superpower** is a state with a dominant position in the international system, capable of exerting its influence (economic, cultural, political) at the global scale.

heavily in its naval capabilities. Its aim is to establish a blue-water navy (one capable of operating away from its home ports and across the deep oceans). China is extending its influence throughout the Indian Ocean, investing in port facilities in countries such as Myanmar, Pakistan and Sri Lanka. This has raised tensions with India, itself an emerging naval power.

The South China Sea: a marine conflict zone

Tensions existed over territorial claims in the South China Sea for centuries and occasionally armed conflict has occurred. There are numerous rocky outcrops and small coral islands that various countries have claimed. The Chinese have been very active in occupying some of the islands, constructing facilities to service ships and aircraft. Although countries such as the Philippines and Vietnam have disputes with China, the strongest military force in the region is the USA. Strategically, the South China Sea is very important, not only to the USA and Japan but also to Europe, as so much trade passes through the region's ports.

Oceans present hazardous obstacles to human activities

Over the course of history, many people have looked to oceans as a means of escape while others have seen them as opportunities to take wealth from others.

Twenty-first-century piracy and its management

The growth in transoceanic trade has been accompanied by a rise in modern piracy. Shipping attack 'hotspots' occur in the western Indian Ocean and southeast Asia. The presence in these regions of **choke points** means that there is a high density of vessels for pirates to attack.

Additionally, the two 'hotspots' mentioned have some politically unstable nations close by from which pirates can operate relatively freely. Pirates have tended to attack large vessels such as bulk carriers and container ships, holding the vessel, its cargo and crew to ransom.

In the western Indian Ocean the changing seasons impact pirate activity. The stronger winds and rougher seas of the monsoons make it difficult for the pirates who use small, fast boats to come alongside and board their intended target.

Management of this issue has brought together a maritime coalition of navies from ACs and some EDCs. Patrols and shared intelligence have countered the piracy threat at sea. Efforts are also being made towards helping dysfunctional governments function more effectively and to offer opportunities for young men that will divert them from criminal activities on the oceans.

Oceans as escape routes for refugees

Whatever the motives of refugees (economic, political, or fleeing war or religious persecution), ocean routes are seen as means of escape despite the dangers of crossing open water. It is important to appreciate that not all migrants originate from countries with a coastline, but also from land-locked countries.

Exam tip

To write with authority about a marine conflict zone, you will need to give factual details about the current state of play but also be able to set this in the historical context. Research your chosen location in full and keep up to date with developments.

Choke point Point of congestion — in shipping, the coastline geography constrains some shipping routes (e.g. Straits of Hormuz, Straits of Malacca).

Boat refugees in the Mediterranean and Asia

Rising numbers of desperate migrants have taken to crossing the Mediterranean between North Africa and Europe. People traffickers take advantage of refugees' plight, charging large sums of money and filling unseaworthy boats with more people than the boat can safely hold. The Mediterranean has flows of refugees coming from Libya, some having crossed the Sahara first, and from Turkey. Australia is the intended destination of many Vietnamese and Tamils from Sri Lanka. Migrants from Bangladesh and Myanmar head southwards, many ending up in Indonesia, Malaysia or Thailand.

As with piracy, global governance is playing a role in the management of refugees. The United Nations High Commissioner for Refugees (UNHCR) attempts to coordinate the global response. Maritime patrols attempt to intercept boats before they sink, and resettlement programmes aim to prevent refugee camps becoming permanent homes for displaced people.

Summary

- Oceans are a distinctive feature of the Earth, with an underwater landscape.
- Salinity and temperature vary both with depth and from one location to another.
- There is a system of interconnected warm and cold currents, both at the surface and deep within oceans.
- Light and nutrient levels vary with depth and from one place to another, so influencing biodiversity.
- Biological ocean resources can be used in sustainable or unsustainable ways.
- Energy (renewable and non-renewable) and mineral resources are exploited within the oceans and can lead to contested issues.
- Governing the oceans is not straightforward and a significant concern.
- Human activities can pollute oceans, such as via offshore oil exploitation.
- Ocean currents can both disperse and concentrate pollution.
- Climate change is altering the ocean's water and levels, so threatening marine ecosystems and human communities.
- Oceans play a vital role in globalisation.
- Oceans are hazardous spaces for some human activities, and conflict can arise when countries challenge each other in these regions.

Future of food
What is food security and why is it of global significance?

There is a consensus that we produce enough food. However, there is a global mismatch between demand and supply. Hunger and obesity exist across the development spectrum. The future of food will be determined by the efficient functioning of the global food system.

The concept of food security and variations in patterns of food security

Defining food security

The UN's Food and Agriculture Organization (FAO) states that:

Food security exists when all people, at all times, have physical and economic access to sufficient, safe and nutritious food that meets their dietary needs and food preferences for an active and healthy lifestyle.

The World Food Programme (WFP) identifies 'three pillars' of food security.

1 **Availability:** addresses food supply

2 **Access:** relates to household-level access to sufficient food

3 **Utilisation:** the intake of food must result in the body gaining sufficient nutrients and energy

The FAO also recognises 'stability' as a fourth pillar to food security. This is important, because food security is dynamic and can also be either long or short term in nature.

Current trends in global food security

A wide variety of data exist to measure food security, including statistics on global hunger, undernourishment and daily calorie intake, and composite indexes that calculate a single value from a number of measures (for example, the Global Hunger Index (GHI), the Global Food Security Index (GFSI)). Since 2000 Latin America and Eastern Europe have reduced their GHI scores by the largest percentage — approximately 40%.

Variations in food security between and within countries

Table 6 summarises the main variations in food security at a global, national and regional (within countries) scale, using a range of measures.

Table 6 The main variations in food security at global, national and regional scales

Global food security	National food security	Regional food security within countries
■ The global regions of Latin America and the Caribbean, eastern Europe, east and southeast Asia and north Africa have all made good progress in achieving food security (GHI and GFSI, 2015) ■ The GFSI shows North America, western Europe and Australasia as the best-performing global regions.	GHI scores for the best- and worst-performing countries achieving reductions in their GHI score (1990–2013): ■ Best performance: Ghana, Kuwait, Mexico, Thailand, Vietnam ■ Worst performance: Burundi, Eritrea, Iraq, Sudan, Swaziland	Variations in food security within countries happen across the development continuum. Examples include: ■ food insecurity in eastern provinces of China ■ in urban areas in many countries there are food-insecure groups, e.g. among the urban poor in Accra, Ghana and inner-city districts, such as the Bronx, New York

Food as an interconnected system

The physical conditions for growing food

Despite technological advances and socioeconomic factors, physical factors exert a major influence over farming methods. The physical conditions required to grow food include air, climate, soil and water (see Table 7).

Exam tip

It is important that you develop the skills to describe the pattern of food security shown by a range of statistical sources and data presentation methods. Your data should also be based on current trends.

Table 7 The physical conditions required to grow food

Temperature	Light	Water	Air	Soil
■ Crops can grow at below optimum temperature, but temperatures that are too high or low result in reduced yields. ■ Tropical crops (e.g. rice) require temperatures between 16°C and 27°C. Temperate crops (e.g. wheat) grow at between 15°C and 20°C.	■ Essential to photosynthesis. ■ Plants differ in light requirements — both light intensity and duration are important for growth.	■ Comprises 80% of living plants. ■ Major determinant of crop productivity and quality. ■ Essential to seed germination and crop growth. ■ Used in photosynthesis to produce sugars from light energy. Acts as a solvent and means of transport for sugars throughout plant.	■ Oxygen and carbon dioxide required for photosynthesis. ■ Plants also require oxygen for respiration, to carry out water and nutrient uptake.	■ Mixture of minerals and organic matter in which roots develop. ■ Plants absorb essential nutrients (nitrogen, phosphorous, potassium, calcium) through roots.

Feeding the world is a complex system

Agricultural systems in different parts of the world are increasingly interconnected to form the 'global food system'. It is a complex network, which includes the production, harvest, processing, transport and consumption of food, and the disposal of waste. The global system does not function as an efficient 'whole' — there are areas of food surplus and food insecurity that have emerged across the development continuum. The policies of individual nation-states, trade, and the actions of organisations such as the FAO, the WFP, the World Bank and a range of transnational corporations (TNCs) influence the system.

Variations in food production methods

Food production methods can be classified according to their scale and input of labour (intensive and extensive) and capital (subsistence and commercial) (see Table 8).

Table 8 Food production methods

Arable and pastoral	Subsistence and commercial	Shifting and sedentary	Extensive and intensive
Arable: the growing of food crops, often on fairly level, well-drained soils of good quality. **Pastoral:** the raising of livestock, often in areas unsuitable for arable farming (too cool, wet or dry, or too steep). Soils often have limited fertility. Livestock farming is sustainable only when the carrying capacity of the area is not exceeded.	**Subsistence:** provision of food by farmers for their own and the local community's consumption. Subsistence farmers are vulnerable to food shortages due to the lack of capital and other entitlements. **Commercial:** farming for profit, often on a large scale with high capital inputs.	**Shifting cultivation:** confined to a few isolated places with low population density, large areas of land and limited food demands (e.g. indigenous groups in tropical rainforests). The system, which is sustainable at low population densities, is essentially a rotation of fields rather than of crops. **Sedentary:** farmers remain in one place and cultivate the same land year after year.	**Extensive:** large-scale commercial farming. Inputs of labour and capital are small in relation to the area farmed. Yields per hectare are low but yields per capita are high. **Intensive:** small-scale with high labour and/or capital inputs and high yields per hectare.
Examples: **Arable farming:** the Nile Valley, the Great Plains **Pastoral farming:** hill sheep farming in Wales, nomadic herding in the Sahel	Examples: **Subsistence:** wet-rice farming in India **Commercial:** cattle ranching in South America, oil palm plantations in Malaysia	Examples: **Shifting cultivation:** the Amazon Basin and Indo-Malayan rainforest **Sedentary:** dairy and arable farming in the UK	Examples: **Extensive farming:** cereal farming on the Canadian Prairies **Intensive:** horticulture in the Netherlands

Globalisation is changing the food industry

The influence of globalisation on the food industry

Food distribution and consumption has become increasingly global, despite the fact that production is a local process determined by factors such as climate, soil and growing season.

- Developments in transport, communications and refrigeration technology have enabled the development of global food chains. Perishable food products can now be transported over long distances.
- Complex transnational food production networks now exist between producers and consumers.
- Food demand has changed as incomes rise through economic growth and populations become more urbanised.
- Food tastes are changing across the world, especially in the affluent markets in North America, Europe and parts of Asia.
- A wide variety of foods is now available from across the globe, with all-year-round provision.

Issues from globalisation of the food industry

Food miles

Due to a combination of changing tastes and improved transportation and technology, food now travels over long distances from producer to consumer (measured in food miles). This is often by air transport, which raises environmental concerns over greenhouse gas emissions. In some instances, however, transporting food can be less environmentally damaging than the out-of-season food production in greenhouses, which impacts levels of energy use.

Inequality between TNCs and small suppliers

The production, distribution and consumption of food have become part of a global industry dominated by TNCs, **agribusinesses** and large food retailers. TNCs often control the terms of farmers' participation in food production. They tend to favour large, capital-intensive operations, leaving small producers disadvantaged and marginalised. In addition, the scale of TNC involvement can make it difficult for national governments to regulate their own food systems and promote the interests of small-scale farmers.

Obesity

As countries develop and affluence increases, consumption of food shifts from cereals to more expensive foods such as meat and dairy products. Development and urbanisation also lead to an increase in fast-food outlets. Weight gain and obesity can result from both of these dietary changes. The rising middles classes in EDCs such as China and Brazil have been significantly affected.

Price crises

Global food prices are volatile and vulnerable to price shocks (an unpredictable and unexpected change in price). Many price increases are due to a sudden shortage in supply as a result of changing weather patterns, but transport, fuel cost and natural hazards can also be responsible.

Exam tip

You should be able to refer to located examples of a range of food production methods.

Knowledge check 17

What factors have led to the globalisation of the food industry?

Agribusiness A large-scale farming practice run on business lines.

Exam tip

It is important to remember that issues created by the globalisation of the food industry impact on countries across the development continuum in some way.

Opportunities from globalisation of the food industry

Technological innovation

There have been significant technological advances in food production. However, it remains a challenge to ensure that these advances are shared between all farmers and countries at varying stages of development.

Technological advances have much to offer in extreme environments, where biotechnology can produce plants that are able to withstand harsh conditions such as drought. Innovation need not always be large scale and high cost — an alternative is the use of **appropriate technology**, such as 'fertiliser deep displacement', where farmers plant slow-release nitrogen briquettes deep in the soil. Mobile phone applications are also an important technological development with widespread potential.

Short-term food relief

Globalisation has facilitated the provision of food aid. Types of international food aid include: bilateral (between two countries), multilateral (provided by a number of countries and agencies) and non-government aid.

Consumer choice

Global food trade has meant that many consumers now have access to a wide range of products through retail outlets and online. Many food brands have global appeal, such as Coca-Cola. Retail giants such as Carrefour have extended consumer choice through multinational locations.

What are the causes of inequality in global food security?

A number of interrelated factors influence food security

Physical, social, economic and political factors can all influence food security.

Physical factors

Geology

Soil forms from the breakdown of rock. This rock, or 'parent material', determines key soil characteristics such as depth, texture, permeability, mineral content and colour.

Soil

The key factors that influence soil formation include the following.

- **Geology/parent material:** determines the structure, texture, mineral content and colour of soils.
- **Time:** it can take anything between 400 and 1,000 years for soil of a few millimetres in depth to form. Agriculture can be practised on newly deposited material, e.g. alluvium.
- **Climate:** the pattern of global soil types corresponds closely to climatic factors.
- **Relief:** at higher altitudes it is colder and so soils take longer to develop. A south-facing aspect will have warmer temperatures, which affects soil formation and water content, while slope affects soil depth and drainage.

Appropriate technology Small scale, can be managed locally and often uses skills available in the local community.

Exam tip

You should be able to refer to a named example of short-term food aid and be aware that food aid brings both opportunities and costs (see also Future of food section, page 73).

Knowledge check 18

Why has obesity become a global issue?

Exam tip

You should have a general understanding of the global distribution of the main farming types.

- **Organisms:** microorganisms such as bacteria are active in the nutrient cycle within soils. Larger organisms, such as worms, mix and aerate the soil.

The interaction of these factors forms a mixture of mineral and organic matter, or soil. Soil covers the Earth's crust in a thin layer and acts as a medium for plant growth. Suitability for agriculture depends on key characteristics of:

- structure (the way soil particles are bound together)
- texture (the size of mineral particles in the soil)
- mineral content (minerals play a vital role in soil fertility and therefore it is essential that mineral stores are allowed to replenish)
- organic matter (decomposed organic matter increases the nutrient supply in soils — in natural ecosystems these can be recycled, but in agro-ecosystems constant harvesting removes nutrients)

Soil characteristics determine farming types and humans often modify these in order to improve their farming potential. If not properly managed, this can lead to a decline in the soil, known as **soil degradation**.

Length of growing season

Each type of crop has a minimum threshold temperature for growth and a growing season of a specific length (for example, some wheat varieties need 90 days for growth). Increases in latitude and altitude reduce the length of the growing season due to climatic factors.

Social, economic and political factors

Land ownership systems

This refers to the ownership rights that farmers have over their land. There are many different types of land ownership and they vary considerably across different countries according to political and economic factors. Land ownership types include the following:

- **Owner-occupiers:** ownership by private individuals, commercial enterprises or governments (sometimes foreign)
- **Tenants:** involves a payment of rent either in money or share of the harvest
- **Landless labourers:** farm workers have no ownership rights

Land ownership impacts food security as it can affect productivity and decision making, and the landowner usually determines the distribution of the harvest.

Competition

Competition exists in a range of contexts within the food supply chain. The following are the two main forms of competition.

- **Competition in food markets:** the growing dominance of agribusinesses and TNCs in global food supply has reduced competition. A lack of competition means that these large organisations determine prices, and small producers and consumers lose out.
- **Competition for scarce resources:** there is increasing global competition for essential agricultural resources such as land, water and energy.

Soil degradation The decline of soil quality caused by its improper use.

Knowledge check 19

How does climate affect soil characteristics?

Exam tip

Make sure that you can relate the social, economic and political factors affecting food security to named examples studied in class

Exam tip

You should be able to refer to contrasting examples of land ownership across various locations.

Land grabbing

The process of land grabbing involves two groups of countries: the target (poorer) country and the investor (richer) country. Investor countries may either have land or water constraints, but high levels of capital (Saudi Arabia), or they may have specific food security issues, such as very large populations (China).

There are a series of benefits for the target countries, which include:

- an increase in agricultural employment
- improvements in rural infrastructure
- development of new agricultural technologies
- creation of local food surpluses

But there are also costs, which include displacement of rural populations, unequal power relations and growing food insecurity.

Theoretical positions on food security

Thomas Malthus (1798)

Malthus's theory was that an optimum population exists in relation to food supply and that an increase beyond this will lead to 'war, famine and disease'. His theories have gained recent support among demographers known as 'neo-Malthusians', based on recent evidence of famines, wars and water security. Malthus stated that in the absence of checks (for example, famine and war), human population would grow at a geometric rate: 1, 2, 4, 8, 16 and so on, doubling every 25 years. Food supply, at best, can only increase at an arithmetic rate: 1, 2, 3, 4 and so on, and is therefore a check on population growth. Given there are limits to the amount of food a country can produce, Malthus suggested preventative measures to limit population growth (for example, abstinence from marriage). Since Malthus' theory was put forward, food production has increased through the use of:

- high yield variety (HYV) seeds/crops
- new foods such as soya
- agrochemicals
- land acquisition, e.g. drainage of wetlands

Ester Boserup (1965)

Boserup believed that countries have the resources, knowledge and technology to increase food supply in response to growth in population, and that population growth is needed to trigger such advancements.

Case study

For this part of the specification you are required to have a case study of **one place** to illustrate how human and physical factors are/have combined to cause issues with food security. Your case study must illustrate:

- the physical challenges to food security presented by the characteristics of key physical conditions such as climate, soil and water
- the human challenges specific to the chosen example — these may relate to government policy, land ownership or the impacts of processes such as globalisation, and can be economic, social, cultural or political in nature
- how these challenges have combined to cause specific food security issues
- an overview of possible future improvements to food security

Exam tip

Make sure that you are able to 'evaluate' the pros and cons of land grabbing and how it affects food security for both the target and investor country. Illustrate your argument with reference to examples.

High Yield Variety (HYV) seeds/crops Developed to improve food supplies and reduce famine in developing countries. They can produce up to ten times more crops than regular seeds on the same area of land.

Agrochemicals Substances used to help manage agricultural ecosystems, e.g. fertilisers, agents to control soil pH, pesticides and chemicals used in animal husbandry (e.g. antibiotics and hormones).

Knowledge check 20

Why have Malthus's theories gained renewed interest?

What are the threats to global food security?

Risks to food security can be identified to highlight the most vulnerable societies

The UN recognised the Right to Food in the Declaration of Human Rights in 1948, yet there are many regions around the world where this right is facing increasing risk

Regions, countries and people whose food security is most at risk across the development spectrum

Regions

There has been food security progress in the regions of the Caucasus and central Asia, eastern Asia, Latin America and northern Africa. There remains very slow progress in southern Asia and sub-Saharan Africa.

Countries

Food security remains fragile in some countries, despite good progress in recent times. Countries in this position include Algeria, Egypt, Morocco and Tunisia.

In other countries, natural and human disasters and political instability have contributed to crisis situations. Countries in this position include Iraq, Syria and Yemen.

People

In 2015, FAO statistics showed that 795 million people were hungry and did not gain enough energy from their diet — 98% of these people lived in LIDCs and included distinct groups of people.

- **Rural dwellers:** 75% of hungry people live in rural areas, mainly in Africa and Asia. They are mostly people dependent on agriculture with no other source of income.
- **Farmers:** half of the world's hungry people are from small-scale communities farming on marginal land
- **Children:** an estimated 146 million children in LIDCs suffer from acute or chronic hunger.
- **Women:** evidence shows that hunger affects women more than it does men. This has further impacts on infant mortality and low birth-weight children.

Pinch points where food security is at risk

Pinch points are places in the global food supply chain where disruption occurs. Globalisation means that food supply chains are increasingly long and complex. Pinch points can develop due to a range of causes, which may be political, environmental, economic or technological in nature. Recent examples include the 2012 UK fuel strikes and the 2010 Iceland volcanic ash cloud.

Exam tip

The global pattern of food insecurity is complex and ever changing. Make sure that you have up-to-date examples of regions and countries where food security is at risk.

Knowledge check 21

At what points in the food supply chain can pinch points create disruption?

The causes of desertification and risks to food security

Desertification is the outcome of persistent land degradation and the reduction in agricultural capacity due to the overexploitation of resources and natural processes, such as drought. In extreme cases this leads to desert-like conditions. It frequently occurs in dryland environments and the scale of the problem means that it is one of the greatest environmental challenges of our time, affecting 74% of the poorest people in the world. The risk to food security from desertification is acute in Djibouti, Ethiopia, Kenya, the Sahel and Somalia.

Causes of desertification

The causes of desertification include environmental, social, political and economic factors, which are interlinked (see Table 9).

Table 9 The human and physical causes of desertification

Human causes of desertification	Physical causes of desertification
■ **Poverty:** lack of farming investment and money forces people to farm any available land, pushing them onto marginal land such as steep slopes, which then becomes overused. ■ **Changing farming practices:** including overgrazing and expansion of cropped areas. New, more intensive agricultural systems deplete soil nutrients as farmers abandon fallow periods and crop rotation. Soils are often already fragile on marginal land. ■ **High demand for irrigation water:** irrigation is a human response to water shortages. Water scarcity causes crops to die and encourages poor farming practices. Irrigation projects are also poorly managed, unsuitable and underfunded. ■ **Demand for fuelwood:** as land is cleared for fuelwood, the soil is left exposed to wind and water erosion. ■ **Political and economic instability:** political instability sometimes leads people to stay in the same areas and land becomes overused. It also leads to displacement of people or individuals (male outmigration), widening the problem and, in the case of male outmigration, leaving women to undertake intensive farming close to the home.	■ **Climate change:** increased periodic drought and changing rainfall patterns damage animal habitats and soil quality. Crops die, farming practices change and land degradation intensifies. ■ **Soil erosion by wind and water:** this removes the top layer of nutrient-rich soil. ■ **Salinisation:** Few plants are salt tolerant.

The increased risks to food security

Desertification has severe impacts on the functioning of ecosystems as it directly affects the quality of soil, which is the basis for plant growth. This triggers a range of further complex, interrelated and wide-ranging impacts on food security. They can be divided into categories of environmental, economic and sociocultural impacts (see Table 10).

Table 10 The environmental, economic and sociocultural impacts of desertification

Environmental	Economic	Social and cultural
■ Continual cropping and fewer fallow periods exhaust soil nutrients and decrease fertility, leading to food shortages. ■ Soil nutrients are lost through wind and water erosion, reducing yields. ■ Loss of biodiversity as vegetation is removed — this impacts the functioning of the ecosystem. ■ Increased dust formation, which can affect cloud formation and rainfall patterns, which in turn increases vulnerability to fluctuating harvests.	■ Reduced availability of fuelwood leads to increased purchase of kerosene, with health issues such as sore eyes and respiratory problems. ■ Food shortages, with a risk of growing dependency on food aid. ■ Reduced income from traditional economies such as pastoralism and cultivation of crops. ■ Widespread rural poverty, leading to food security issues, as people cannot afford food.	■ Dryland populations are often socially and politically marginalised due to poverty and remoteness. ■ Forced migration, due to scarcity of productive farmland and hunger. ■ Increased male outmigration, leaving women to undertake responsibility for farming and water management. ■ Loss of traditional knowledge and skills, leading to falling yields.

The food system is vulnerable to shocks that can impact food security

How extreme weather events can affect food production

Extreme weather events include El Niño, heatwaves, droughts, wildfires, floods, hurricanes and tornadoes. They are all becoming more frequent and more severe around the world. The link between global warming and increased severe weather can be explained as follows:

Warming atmosphere ⇒ more evaporation from the oceans ⇒ more water vapour in the atmosphere ⇒ water vapour helps the Earth to hold onto heat energy from the sun ⇒ further global warming and severe weather events such as heatwaves and prolonged periods of rainfall

There are several key links between climate change and food production, which will vary in their impact across the globe:
- Food production is a driver of climate change through the release of carbon dioxide by deforestation, land clearance and methane-intensive crop and livestock farming.
- Climate change will lead to extended growing seasons in some areas and a reduced season in others.
- Wildfires, floods, storms and drought directly impact food production.
- Farming affects carbon storage through the input and output of organic matter.
- Extreme weather events can impact food distribution by affecting transport infrastructures, and this can exacerbate food security issues.

Exam tip

You should know the general global pattern of the potential impact of climate change on food production and the resulting vulnerability to food insecurity. Appropriate use of examples can give your answers authority.

How water scarcity can exacerbate food production issues

Agriculture accounts for 68% of water drawn from rivers, lakes and aquifers. Of the water available, up to 60% is lost due to poor irrigation systems and high levels of evapotranspiration. Water scarcity causes crops to die and encourages poor farming and irrigation practices.

In countries where climatic and/or human factors mean that fresh water is scarce, food production has to be carefully adapted. Such environments include areas of intense drought or cold, or with high rates of evapotranspiration. Human factors include limited access to water due to a lack of resources and technology, or the use of poor water management techniques.

Exam tip

Always give a balanced response to climate change questions regarding the impact on agricultural production, as some regions will have higher yields due to a warming climate.

Many countries across the development spectrum are trying to find more efficient means of using water in agriculture. One such measure relates to the transfer of **virtual water**. In water-scarce countries, importing water-intensive food products (those that require a high input of water, such as lettuce) relieves pressure on domestic water resources. Another measure includes countries offering farmland to water-scarce countries, but this can then compromise food security in the host country.

Virtual water The volume of fresh water needed to make a product, measured at the place of production.

How tectonic hazards can influence food production and distribution

Tectonic hazards can have a positive impact on food production, such as the high levels of fertility of volcanic soils. However, in terms of shocks that can impact food security, there are negative impacts on both production and distribution (see Table 11).

Table 11 How volcanic eruptions and earthquakes can affect food security

Volcanoes	Earthquakes
■ Volcanic ash fall destroys pasture land. ■ Volcanic ash can increase the sulphur levels in soils and lower pH to levels where plants cannot survive. ■ Ash fall can adhere to fruit skins and either destroys crops or it is too expensive to clean them for sale to foreign markets with rigid standards.	■ Transport and food distribution may be disrupted by cracks from earthquake activity. ■ Food stocks may be destroyed. ■ Livestock may be killed. ■ Irrigation systems may be damaged.

Case study

For this part of the specification you are required to have a case study of **one indigenous farming technique** in an extreme environment. Your case study must illustrate:

■ the physical conditions of the environment, including ecosystems, terrain and climate
■ food production methods used by indigenous people in the environment
■ threats to the indigenous group's food security

How do food production and security issues impact people and the physical environment?

Imbalance in the global food system has physical and human impacts

Reasons for recent increases in global food supply have been higher yields per unit of land, crop intensification and an increase in the amount of land being farmed globally.

However, this has also resulted in several impacts on both the physical environment and humans.

Physical impacts of attempts to increase food production

A summary of the possible impacts of food production on the physical environment is shown in Figure 8.

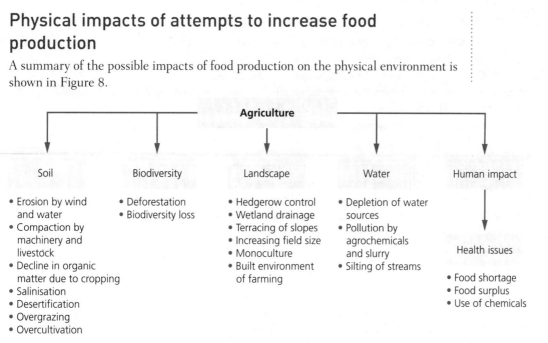

Figure 8 Summary of the environmental impacts of food production

Irrigation and salinisation

Salinisation is a natural process, often made worse by human activity. The process can be explained as follows:

Low precipitation and high evaporation $\Rightarrow$ salts in the soil are brought to the surface $\Rightarrow$ plants intake water but leave salts behind $\Rightarrow$ a salt layer is left, which is toxic to plants and can inhibit water absorption and directly affect plant growth $\Rightarrow$ land is consequently unusable for agriculture.

Poor irrigation techniques can intensify salinisation when water is brought to land that is naturally dry as:

- irrigation can lead to increased rates of groundwater recharge, causing water tables to rise — this brings salts to the plant root zone, which affects both plant growth and soil structure
- water used to irrigate crops evaporates quickly in very dry conditions
- inefficient irrigation with high levels of leakage raises water tables and increases the risk of salinity
- excessive withdrawal of underground water in coastal areas leads to infiltration of saline water into fresh supplies

The impacts of irrigation salinity include:

- reduced agricultural productivity
- reduced income for farmers
- a decline in water quality, potentially impacting livestock and future irrigation
- weakened structure, nutrient loss and increased erosion in soils
- a decline in natural vegetation, with the consequent loss of habitat for some species

> **Knowledge check 23**
>
> Explain the process of salinisation.

Deforestation and the impacts on biodiversity

Increasingly, natural forest is being converted into agricultural land. This is especially true in developing countries, where in some regions (for example, the tropics) this can account for up to 80% of new agricultural land. In addition, the increase of large agribusinesses taking over from small farmers has meant that the practice is now on a much larger scale. LIDCs without valuable resources such as oil and gas depend on agriculture to support the economy. In such economies, conversion of forest to agricultural land continues at a rapid pace.

Cropping intensity and higher yields may help the situation in some countries but there are estimates that across the developing world, cultivated land will increase by almost 50% by 2050, with 66% of new land coming from deforestation. Species richness is closely related to the occurrence of wild habitats. Deforestation to acquire agricultural land, hedgerow removal and grazing and drainage of wetlands all reduce natural habitats and biodiversity.

Changing landscapes

Due to the rise of modern industrial farming methods and the need to increase production, there has been a considerable impact on natural landscapes. Driving forces include farming practices such as deforestation, climate change resulting in desertification, and changes in national, regional and local farming policies.

Specific landscape changes in the UK include:

- growing field sizes
- conversion of natural habitats for food production, e.g. wetlands, small woodlands, moorlands
- Monoculture, which leads to a less varied landscape
- More 'built environment' features, e.g. polytunnels, silos, greenhouses, solar panels

Water quality from agrochemicals

Agrochemicals such as herbicides, insecticides and pesticides have become increasingly important in crop production. Intensive use of these chemicals leads to environmental problems, such as contamination of soil and groundwater. When applied, only around 15% of the preparation hits the target — the rest is distributed in the soil and air (pesticide drift).

Agrochemicals in soil can move from the surface when they are dissolved in runoff water, or when they percolate down through the soil. Those that have infiltrated the soil will eventually reach groundwater sources through the process of leaching. Emerging economies are at particular risk because there is less monitoring and control of agrochemical use than in developed economies.

Leaching Soluble materials draining away in soil.

Table 12 expands on the impacts of pesticides and fertilisers and manure on both surface and groundwater.

Table 12 Agricultural impacts on water quality

Agricultural activity	Impacts	
	Surface water	Groundwater
Fertilising	■ Runoff of nutrients, especially phosphorus, leading to **eutrophication**, causing taste and odour in public water supply, excess algae growth (algal bloom), and leading to deoxygenation of water and fish kills	■ Leaching of nitrate to groundwater — excessive levels are a threat to public health
Manure spreading	■ Carried out as a fertiliser activity, spreading on frozen ground results in high levels of contamination of receiving waters by pathogens, metals, phosphorus and nitrogen, leading to eutrophication and potential contamination	■ Contamination of groundwater, especially by nitrogen, which is a soluble compound that can easily leach from soil into deep aquifers by percolation
Pesticides	■ Runoff of pesticides leads to contamination of surface water and biota, dysfunction of the ecological system in surface waters by loss of top predators due to growth inhibition and reproductive failure, public health impacts from eating contaminated fish ■ Pesticides are carried as dust by wind over very long distances and contaminate aquatic systems thousands of miles away (e.g. tropical/subtropical pesticides found in Arctic mammals)	■ Some pesticides may leach into groundwater, causing human health problems from contaminated wells

Eutrophication
Excessive richness of nutrients in a body of water, frequently due to runoff from the land, which causes a dense growth of plant life.

Knowledge check 24

How does the addition of fertiliser to agricultural land alter freshwater ecosystems?

Case study

For this part of the specification you are required to have a case study of how food production methods have impacted **one physical environment**. Your case study must illustrate:

■ the specific short- and long-term impacts on the environment

Exam tip

Make sure you have a full understanding of the nutrition spectrum, which sets out a scale of health issues associated with under and over nutrition.

How food security issues impact people

Health issues associated with food shortages

Food security refers not only to access to sufficient food but also to safe and nutritious food that meets dietary needs and helps maintain a healthy life. Nearly 30% of the world's population suffers from **malnutrition**, where daily calorie intake is low and leads to decreasing health issues. However, increasingly a different set of health issues is associated with excessive calorie intake. The health issues associated with high and low food consumption are set out in the 'nutrition spectrum'. Geographically, there are high levels of food consumption in North America, Europe, parts of North Africa, Australia and New Zealand, and low levels of consumption in most of Africa, and central and northern Asia.

Malnutrition The shortage of proteins and essential vitamins caused by an unbalanced diet.

Health issues associated with food surpluses and poor diet

Obesity

The number of overweight people now exceeds those underweight. Obesity affects several developing nations, for example Colombia (41%) and Brazil (36%). Such is the spread that we now use the term 'globesity'.

The cause of obesity is an energy imbalance between calories consumed and calories expended. Also, there has been increased intake of energy-dense foods high in fats and sugars and low in vitamins and minerals, such as processed foods, ready-meals and fast foods. Health consequences of obesity include non-communicable diseases such as diabetes, cardiovascular diseases (CVDs) and some cancers. Children with poor diet and a lack of exercise who become overweight risk health consequences such as fractures and early markers of CVD.

Responses to obesity exist at a range of levels:

- personal level — adapting to a healthier lifestyle with more exercise and reduced calorie intake
- government level — ensuring that a healthy diet is affordable and accessible to all, and that the food industry has responsibility to reduce unhealthy components of processed foods and set attainable price levels

Harmful impacts of chemicals and pesticides

The use of insecticides, pesticides and chemicals has increased globally in order to control harmful crop diseases that threaten yields and to increase overall food production. Monitoring and management of the impacts of **agrochemicals** has led to a range of health scares.

In ACs such as the UK, organisations including the Food Standards Agency (FSA) exist to protect public health regarding food safety issues. In EDCs and LIDCs there is less regulation, which is of growing concern.

> ### Case study
>
> For this part of the specification you are required to have a case study of **two places** at contrasting levels of economic development. Your case studies must illustrate the implications of poor food security on the lives of people living in those places.
>
> This should acknowledge the broad definition of food security, as it refers to not only adequate supply of food but also both quality and affordability of food (a supply of safe and nutritious food for a healthy life). Remember that food security issues affect countries at contrasting levels of the development spectrum.

Is there hope for the future of food?

Food is a geopolitical commodity

Food is a geopolitical commodity and a number of key players will continue to influence the global food system (see Table 14 on page 74).

Knowledge check 25

How can food nutrition be improved?

Agrochemicals Any substance used to help manage an agricultural ecosystem, e.g. fertilisers and pesticides.

Exam tip

The global food problem is not caused by scarcity of food but rather the distribution of food resources and access to markets, technology and basic resources such as land and water.

The geopolitics of food

A series of events such as the global economic recession, food supply shocks, civil unrest, food riots and concern over the long-term food supply as a result of global warming have meant that the food supply is increasingly affecting and being influenced by political decisions and events. Examples include countries defending their food supply with food export bans and acquiring land for food production beyond their borders (for example, the Gulf States buying land in Africa).

The opportunities between countries to ensure food security

Agricultural trading policies

Trade in food is needed to ensure global food security. Agriculture accounts for more than one-third of export earnings in 50 developing countries. Trade in food has increased five- fold over the past five decades. However, the export subsidies and import tariffs of developed countries mean that some poor countries are unable to compete in international markets. Different types of food trade agreements exist (see Table 13).

Table 13 Different types of food trade agreements

Trading blocs	Multinational agreements	Bilateral trade agreements
An agreement between a number of countries to promote free trade among its members. Tariffs are imposed on the imports from non-member states. Example: the EU	Several countries engage in a trading relationship with a third party. Example: ACP nations being given free trade access to EU markets	A trade agreement between two political entities that has mutual benefits and is legally binding. Example: Sainsbury's trade agreement with St Lucia for fair trade bananas

The World Trade Organization

The World Trade Organization (WTO) was set up in 1995 as a continuation of the General Agreement on Tariffs and Trade (GATT). Based in Geneva, Switzerland as of 2015, the WTO has 161 members representing 97% of world trade. Its main role is to provide a forum for governments to negotiate trade agreements and settle trade disputes. Some criticise the WTO for not doing enough to achieve equitable global food distribution.

Appropriate aid

Food aid has its benefits, particularly in crisis situations such as Syria's civil conflict and the 2015 Nepal earthquake. Types of aid include the following:

- **Programme food aid:** the transfer of food from one government to another as a form of economic support. May be used by donors for surplus disposal, to achieve political objectives or to expand markets overseas.
- **Project food aid:** provided for hunger-related development, disaster relief or nutrition programmes. Most project food aid is directed through multilateral agencies, e.g. the WFP.
- **Emergency or relief food aid:** intended for direct, free distribution to people facing famine or an acute food shortage because of natural or human-made disasters.

Knowledge check 26

Explain how food has become a geopolitical commodity.

Exam tip

It is important to understand how unfair trade deals can impact farmers in LIDCs with a high level of dependency on farming income. Examples include paying unstable prices to LIDCs due to disease and climate, or using the best land for cash crops for export, leaving only marginal land for subsistence farmers.

Knowledge check 27

Why do some criticise the WTO's role in food security?

Criticisms of food aid include the following.

- Donor-driven food aid centres are a means of 'dumping' surplus food from ACs.
- Food aid dependency for the recipient countries could be a long-term outcome of this type of aid. Large quantities of food aid can swamp local markets and drive down prices, reducing the income of indigenous farmers.

The alternatives to food aid include fair trade (which includes principles such as fair prices and good working conditions), appropriate technology and small-scale projects to promote the recovery of farming. For example, the MERET Ethiopia programme helps farmers reclaim degraded land using simple techniques such as terracing hillsides to prevent soil erosion (see www.wfp.org/disaster-risk-reduction/meret).

> **Knowledge check 28**
>
> Outline the costs and benefits of the involvement of TNCs in the food industry.

The roles and responsibilities of key players in influencing the global food system

Table 14 Key players' influences on the global food system

Agribusiness	TNCs	Food retailers	Fair trade organisations
<ul><li>A large-scale farming practice run on business lines.</li><li>Involvement extends through the food supply chain, particularly in production, processing and distribution.</li><li>Criticism that in the pursuit of profit, environmental issues have been compromised, especially through their use of agrochemicals and hormone growth promoters.</li></ul>	<ul><li>A very large company with factories and offices in more than one country.</li><li>They often specialise in 'downstream activities' such as processing and distribution of food, e.g. Kraft.</li></ul>	<ul><li>There is a dominance of global food retailers, e.g. supermarkets control 60% of food retailing in Latin America.</li><li>Local traders are unable to compete against them, e.g. Tesco, Carrefour.</li></ul>	<ul><li>The World Fair Trade Organization exists to promote fair trade and greater equality in international trade.</li></ul>

Strategies that exist to ensure and improve food security

Approaches to increasing food security

A range of physical and human factors threaten food security, including climate change, natural hazards, land degradation, financial crises, unfair trade deals, competition and price rises. A range of approaches also exist to ensure and improve long-term food security:

Short-term relief

This would include methods such as food aid administered in a crisis situation to alleviate immediate risks to food supply. Long-term dependency on food aid can cause problems.

Capacity building

This refers to the capability of countries and communities to build a resilient food supply system that can withstand threats such as those listed above. It can be achieved by economic development and access to fair trade deals, for example.

Long-term system redesign

This approach refers to a country's long-term strategic plan to achieve and maintain food security, from securing long-term supply to educating a population in lifestyle choices to ensure they eat a healthy, balanced and nutritious diet.

The effectiveness and sustainability of techniques to improve food security

Examples of the approaches outlined above are summarised in Figure 9.

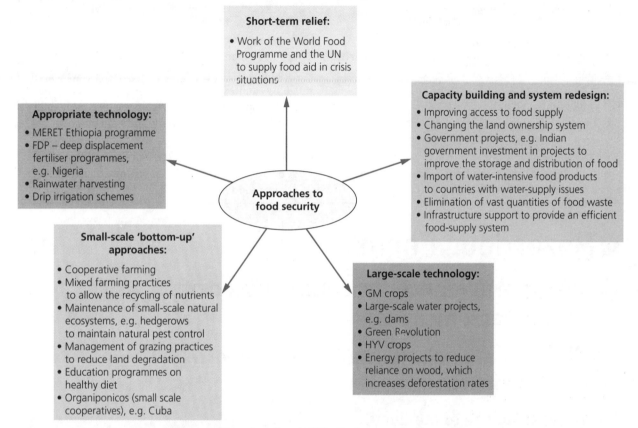

Short-term relief:
- Work of the World Food Programme and the UN to supply food aid in crisis situations

Capacity building and system redesign:
- Improving access to food supply
- Changing the land ownership system
- Government projects, e.g. Indian government investment in projects to improve the storage and distribution of food
- Import of water-intensive food products to countries with water-supply issues
- Elimination of vast quantities of food waste
- Infrastructure support to provide an efficient food-supply system

Appropriate technology:
- MERET Ethiopia programme
- FDP – deep displacement fertiliser programmes, e.g. Nigeria
- Rainwater harvesting
- Drip irrigation schemes

Approaches to food security

Small-scale 'bottom-up' approaches:
- Cooperative farming
- Mixed farming practices to allow the recycling of nutrients
- Maintenance of small-scale natural ecosystems, e.g. hedgerows to maintain natural pest control
- Management of grazing practices to reduce land degradation
- Education programmes on healthy diet
- Organiponicos (small scale cooperatives), e.g. Cuba

Large-scale technology:
- GM crops
- Large-scale water projects, e.g. dams
- Green Revolution
- HYV crops
- Energy projects to reduce reliance on wood, which increases deforestation rates

Figure 9 Examples of techniques to improve food security

Case study

For this part of the specification you are required to have a case study of **two contrasting places** at different levels of development. Your case study must illustrate:

- the strategies and techniques that have been used to ensure or improve food security
- how effective have they been
- to what extent the practice is sustainable

Exam tip

Even when not directed to do so by the question, it is good practice to illustrate answers where appropriate with case study material.

Exam tip

It is important not only to be able to refer to a range of examples of strategies to ensure and improve food security, but also to be ready to assess how effective and sustainable they are.

Summary

- Food security is a complex and contested term with distinctive global patterns.
- Globalisation has had wide-ranging impacts on global food security, which includes both opportunities and issues.
- Physical and human factors affect food security.
- Despite increases in food production, food insecurity exists for a range of different people both within and between countries.
- Physical processes that threaten food security on a global scale include desertification and salinisation.
- Extreme weather conditions, climate change and natural hazards all threaten food security.

- Impacts of food security on the physical environment include land and soil degradation, biodiversity reduction, changing landscapes and declining water quality.
- People experience a range of health impacts as a result of food insecurity.
- Food has become a geopolitical commodity.
- Opportunities exist to improve food security, and profit-making organisations have a responsibility to protect the environment and help achieve global food security.
- A variety of approaches to increasing food security exist at a range of scales and over the short and long term.

Hazardous Earth

What is the evidence for continental drift and plate tectonics?

Evidence for the theories of continental drift and plate tectonics

The basic structure of the Earth

The Earth has a concentric structure with three main layers: core, mantle and crust. Each of these layers can be sub-divided (see Table 15).

Table 15 Summary of the Earth's internal structure

Layer	Starting and finishing depths	Characteristics
Crust	Continental: surface to average 35 km, maximum 70 km	Solid rock: mainly granitic, silicon and aluminium (sial), density 2.6–2.7
	Oceanic: surface to between 5 km and 10 km	Solid rock: mainly basaltic, silicon and magnesium (sima), density 3.0
Mantle	Lithosphere: crust to 100 km	Rigid layer
	Asthenosphere: 100 km to 300 km	Semi-molten (easily deformed)
	Mesosphere: 300 km to 2,900 km	Solid material
Core	2,900 km to about 6,400 km	Inner core (solid) Outer core (liquid)

Convection currents are believed to operate within the asthenosphere. Their slow movement means that the solid lithosphere and crust are also moved around the Earth's surface.

Continental drift and the theory of plate tectonics

As the outlines of the continents were accurately mapped, ideas emerged about how the land masses might fit together, for example South America and Africa.

The ideas of Alfred Wegener

In the early twentieth century, Alfred Wegener proposed that the continents slowly drifted about the Earth's surface. Starting in the Carboniferous geological period (250 million years ago), a large single continent, Pangaea, broke up. Eventually the continents as we see them today emerged. Wegner's ideas were fairly widely accepted, but evidence of how the continents might move was not available (see Table 16).

Table 16 Wegener's evidence for continental drift

Geological evidence	Biological evidence
■ The fit of continents such as South America and Africa on either side of the Atlantic. ■ Evidence from about 290 million years ago of the effects of contemporaneous glaciation in southern Africa, Australia, South America, India and Antarctica, suggesting that these land masses were joined at this time, located close to the South Pole. ■ Mountain chains and some rock sequences on either side of oceans show great similarity, e.g. northeast Canada and northern Scotland.	■ Similar fossil brachiopods (marine shellfish) found in Australian and Indian limestones. ■ Similar fossil reptiles found in South America and South Africa. ■ Fossils from rocks younger than the Carboniferous period in places such as Australia and India, showing fewer similarities, suggesting that they followed different evolutionary paths.

Palaeomagnetism

As lava cools, particles of magnetic minerals become orientated in the direction of the Earth's magnetic field. It is now known, through studies in paleomagnetism, that this field reverses periodically, however, it is not known why these reversals occur. A reversal takes about 20,000 years to complete and occurs at various frequencies through geological time.

Strips of rocks with the same magnetic orientation occur symmetrically either side of mid-ocean ridges. Sea-floor spreading occurs as molten material rises up to the ocean bed and 'pushes' older rock away from the ridge.

The age of sea-floor rocks

Once deepwater drilling was possible, sediments were recovered from the ocean bed. These showed that the thickest and oldest sediments were near the continents, with younger deposits further out in the oceans. Dating analysis revealed that no sea-floor rocks were older than 200 million years, which is young compared with the continents. This supported the idea of sea-floor spreading and of continental drift.

Convection currents
Occur due to parts of the asthenosphere becoming very hot, rising and moving sideways when they reach the lithosphere. The hot material cools and sinks back down.

Sea-floor spreading
The lateral movement of new oceanic crust away from a mid-ocean ridge.

However, sea-floor spreading suggests that the Earth should be becoming larger, but this is not the case. Large-scale trenches were discovered in parts of the oceans where rock was being forced down back into the asthenosphere, giving further support to ideas of continental drift and plate tectonics.

The theory of plate tectonics states that the Earth's crust and lithosphere are divided into a number of rigid plates. These can be moved by convection in the asthenosphere. The movement of these plates explains the distribution of earthquakes, volcanoes and their associated landforms. The theory, developed from the 1960s, has revolutionised earth science in terms of knowledge and understanding of landforms and processes.

The global pattern of plates and plate boundaries

Detailed maps of where **seismic** activity is located emerged during the 1960s. Narrow bands of particularly strong seismic activity snake over the Earth's surface both on land and under the sea. In between were vast areas where little seismic activity was recorded. Today, the pattern of seven large plates (e.g. Pacific, Eurasian) and many minor ones (e.g. Philippine) is well known.

There are distinctive features and processes at plate boundaries

Research continues into what happens at plate boundaries. We recognise three basic types of boundary, their differences being the relative movements of the plates and the resulting processes, landforms and landscapes.

1 Divergent (constructive) plate boundaries

Magma rising up from the asthenosphere has the force to push apart plates as convection currents circulate under the lithosphere. Such splits in the crust tend to occur at mid-ocean ridges. These are long chains of mountains, sometimes rising 3,000 m above the ocean bed.

As magma rises the rigid lithosphere and crustal rocks are forced up into a dome. Because they are rigid, the enormous stresses created as the dome rises result in the rocks fracturing, producing **rift valleys**. The resulting earthquakes are frequent but not high energy.

In the North Atlantic, magma production has been so great that a volcanic island has formed — Iceland, which sits on the mid-oceanic ridge. Running across the island is a dramatic rift valley representing the divergent boundary between the North American and Eurasian plates.

Most of the material erupted along divergent boundaries is underwater. As the magma meets sea water, it cools rapidly, forming rounded mounds of lava characteristic of lava erupting underwater (pillow lavas). In some locations, magma superheats sea water that seeps into the rocks. The very hot water rises, causing chemical changes to the basaltic rocks it contacts. Jets of hot water containing a thick 'soup' of minerals blast out from the ocean bed (black smokers). Unique and highly specialised organisms and ecosystems exist at these locations.

Seismic This term means 'of an earthquake', as in seismic energy or seismic waves.

Rift valley A linear valley formed by the sinking down of rocks between fractures or faults. The sides are often steep, the valley floor relatively flat.

Mid-ocean ridges do not run continuously. Every so often **transform faults** cross the ridge, displacing it sideways and breaking it into segments. As these faults slip, earthquakes happen but no volcanic activity occurs.

Rifting away from mid-oceanic ridges

Doming and rifting can take place on land where the crust thins. The crust between Egypt and Turkey has been stretched sufficiently for faulting to occur, and a rift valley or **graben** formed. In places, the valley has sunk below sea level, forming the Red Sea and further north in Israel, the Dead Sea. If the crust continues to thin along this rift, a new divergent boundary may emerge between Africa and the Arabian Peninsula.

2 Convergent (destructive) plate boundaries

Three combinations of plates can converge or come together:

1 oceanic–continental

2 oceanic–oceanic

3 continental–continental

Oceanic–continental plate boundaries

Due to their different densities, when these two types of plate meet, **subduction** occurs. As the oceanic plate descends it forms a very deep **ocean trench**. The angle of descent is usually between 30° and 70°. The oceanic plate is put under great pressure and friction. In the **Benioff zone** much faulting and fracturing of rock occurs releasing vast amounts of seismic energy.

The descending plate also melts and as the molten material is less dense than its surroundings, it rises towards the surface. Sometimes large intrusions of magma are injected into the overlying rocks of the continent, known as plutons. At some locations magma reaches the surface, giving rise to volcanic eruptions.

The sediments carried by oceanic plates are either subducted or crumpled up along with the leading edge of the continent to form mountain chains. Some of these are among the highest and most extensive ranges on Earth, such as the Andes.

Oceanic–oceanic plate boundaries

When two oceanic plates meet, the slightly denser one will be subducted, forming an ocean trench. As the descending plate melts, magma rises to the surface, erupting as a chain of volcanoes known as an island arc, such as the Aleutian Islands in the north Pacific.

Continental–continental plate boundaries

The coming together of two continental plates results in little subduction. Rather the two tend to 'grind' together, releasing much seismic energy. As the two plates converge, rocks are compressed and extensive fold mountain ranges form, such as the Himalayas.

3 Conservative plate boundaries

In some locations the relative movement of plates is essentially one where they slide past each other. As there is no subduction, no volcanic activity results. However, frictional resistance affects the movement, causing the build-up of pressure. This is released as earthquakes.

Transform faults Large-scale features at right angles to a mid-oceanic ridge. They range in length from a few tens of kilometres to several hundred.

Graben The downfaulted section of a rift valley.

Subduction Occurs where the oceanic crust sinks below either a continental or oceanic plate along a destructive boundary (convergent margin).

Ocean trench A long, narrow depression, mostly between 6–11,000 m deep. They are asymmetric in cross-profile, the steeper side on the continent side.

Benioff zone The boundary between a subducting ocean plate and the overriding continental plate.

The impact on the landscape and landforms can be to leave a 'tear' showing where the plates have moved relative to each other. This can disrupt drainage patterns, deflecting river courses. At some places there can also be some slight vertical movement, resulting in a steep, but small (few metres at most), drop.

What are the main hazards generated by volcanic activity?

There is a variety of volcanic activity and resultant landforms and landscapes

Several factors influence the type of volcanic activity occurring at any one location:

- type of plate boundary
- chemistry of the lava and its **viscosity**
- materials erupted, e.g. ash, pumice, lava
- gases produced
- how the eruption takes place

Eruptions can be placed into one of two groups.

1 **Explosive eruptions:** violent due to the build up of pressure within the volcano. Tend to have viscous lava such as andesite, which can block the volcano vent.

2 **Effusive eruptions:** much less violent due to free-flowing basic lava such as basalt.

The causes and features of explosive and effusive eruptions

See Table 17 for the characteristics of these two groups of eruption.

Table 17 Characteristics of explosive and effusive eruptions

	Explosive eruptions	Effusive eruptions
Location	Convergent plate boundaries	Divergent plate boundaries
Type of lava	Rhyolite (more acid), andesite (less acid)	Basalt
Lava characteristics	Acid (high % silica), high viscosity, lower temperature at eruption	Basic (low % silica), low viscosity, higher temperature at eruption
Style of eruption	Violent bursting of gas bubbles when magma reaches surface, highly explosive, vent and top of cone often shattered	Gas bubbles expand freely, limited explosive force
Materials erupted	Gas, dust, ash, lava bombs, tephra	Gas, lava flows
Frequency of eruption	Tend to have long periods with no activity	Tend to be more frequent, an eruption can continue for many months
Shape of volcano	Steep-sided **strato-volcanoes**, **caldera**	Gently sloping sides, **shield volcanoes**, lava plateaux when eruption from multiple fissures

Explosive eruptions

Acid magma does not flow readily, so the vent of a strato-volcano can often be filled with a mass of solidified magma. This allows great pressure to build up when fresh magma rises and violent eruptions occur, sometimes leading to a caldera forming. Inside the cone, small-scale igneous features, **sills** and **dykes**, can form.

Effusive eruptions

Basic lava is free-flowing. It often erupts from multiple fissures and can be produced in vast quantities to form **flood basalts** — the Deccan Plateau extends for some 500,000 km². Shield volcanoes are often associated with divergent boundaries and are therefore mostly underwater. They are vast structures extending horizontally for many tens of kilometres.

Eruptions at hotspots

The Pacific plate has been slowly moving over a **hotspot** that has produced the Hawaiian chain of islands. The older islands, essentially extinct shield volcanoes, have lost their source of magma as the plate has carried them away from the hotspot. Active volcanoes exist on Big Island (Hawaii). Some 30 km off the coast of Hawaii the next island is forming. Currently, the summit of this shield volcano is just under 1,000 m below the ocean surface and eventually it will grow to become the next island in the chain.

Super-volcanoes

A super-volcano is one that erupts more than 1,000 km³ in a single eruption event. It tends to exist as a giant caldera. Evidence of the impacts of these high-magnitude events comes from how far the ash spread and its depth, and from paleobiological assessments of effects on ecosystems such as mass mortalities. The most recent super-volcano event was some 27,000 years ago, at Taupo, North Island, New Zealand. The Yellowstone super-volcano in Wyoming, USA is intensively monitored and researched.

Measuring and assessing volcanic activity

Two key factors need to be investigated:

1 magnitude (amount of material erupted)

2 intensity (speed at which material erupted)

Two volcanoes can erupt the same volume of material but one does so over a long period (months/years), the other in just a few hours. These are very different volcanic events.

The Volcanic Explosivity Index (VEI) combines magnitude and intensity into a single number on a scale from 0 (least explosive) to 8 (most explosive). It is a logarithmic scale, so each increase in number represents a tenfold increase in explosivity. It takes into account factors such as volume of erupted material, height the erupted material reaches and length of eruption in hours. It is not that useful for effusive eruptions but offers a way of assessing the relative impacts volcanoes can have.

Sill Forms when magma intrudes between the rock layers, forming a horizontal or gently dipping sheet of igneous rock.

Dyke Forms as magma pushes up towards the surface through cracks in the rock. Dykes are vertical or steeply dipping sheets of igneous rock.

Flood basalt A very large area of basaltic lava erupted over many hundreds or thousands of years from multiple eruption events.

Hotspots Locations where magma from a particularly active area of the asthenosphere rises up and breaks through the crust. These eruptions are not associated with plate boundaries.

Exam tip

Research examples of the different types of eruptions so that you can use such material when discussing the impacts of volcanic activity.

Volcanic eruptions generate distinctive hazards

Volcanoes become hazards (see Figure 10) when they interact with humans.

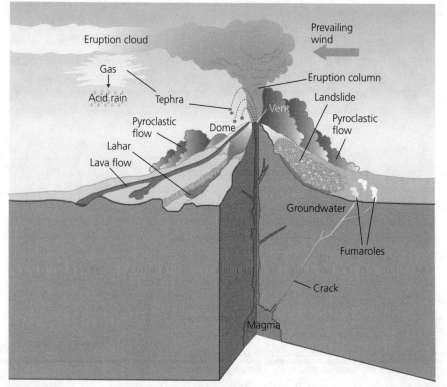

Figure 10 Types of volcanic hazard

- **Lava flows:** basic (e.g. basalt) lava can run for long distances, several kilometres. Acidic lavas (e.g. rhyolite) are thick, rarely flowing far. Everything in the path of a flow will be destroyed, so the main hazards are to infrastructure, property and crops. Few injuries or fatalities result.
- **Pyroclastic flows:** hot gas (500°C+), ash and rock fragments travelling at high speed (100 km/h+) following the shape of the ground. Devastating to anything in their path. Pyroclastic flow from Mount Vesuvius overwhelmed Pompeii.
- **Tephra:** any material ejected from a volcano, ranging from fine ash to large volcanic bombs (> 6 cm across). Potentially very hazardous — ash causes breathing difficulties, smothers crops, collapses buildings through build-up on roofs, and disrupts transport, ground and air.
- **Toxic gases:** wide range of gases ejected during an eruption, e.g. carbon monoxide, carbon dioxide and sulphur dioxide. Can cause death to living things and acid rain, which results in longer-term pollution of soils and water.
- **Lahars:** mud flows often including larger rocks and boulders with the consistency of wet concrete. Can travel at speeds up to 50 km/h. Destroy or bury everything in their path.

- **Floods:** eruptions beneath ice cause rapid melting. Water builds up until it finds an exit and then gushes out in a torrent. In Iceland these are known as jökulhlaup. Erupted material can also block valleys causing flooding if this dam breaks.
- **Tsunami:** violent eruptions of an island volcano can displace vast quantities of water, sending tsunami waves travelling at up to 600 km/h.

Many volcanic hazards pose relatively short-term risks. However, some eruptions eject vast quantities of ash into the upper atmosphere, which has implications for longer-term climatic change. The super-volcano Toba erupted some time between 69,000 and 77,000 years ago. The ash and atmospheric sulphuric acid produced caused global cooling as incoming sunlight was blocked.

Knowledge check 30

Why is lava chemistry an important factor in influencing how dangerous a volcano is?

What are the main hazards generated by seismic activity?

There is a variety of earthquake activity and resultant landforms and landscapes

What is an earthquake?

Earthquakes result from the release of stress within the crustal rocks. They possess a **focus** and an **epicentre**. There tends to be a sequence of foreshocks leading up to the highest energy quake, which is then followed by less energetic aftershocks.

Seismic activity is concentrated in four locations.

1 **Mid-ocean ridges:** tension comes from sea-floor spreading
2 **Ocean trenches and island arcs:** compressive forces associated with subduction
3 **Collision zones:** compressive forces associated with grinding together of plates
4 **Conservative plate margins:** shearing forces associated with irregular movement of plates past each other

In addition, earthquakes occur away from plate margins as stress within the crust can happen anywhere. Old faults, former plate boundaries, the collapse of old mines and the extraction of oil and gas can cause earth tremors.

Focus The point at which the shock waves originate in an earthquake.

Epicentre Location on the Earth's surface immediately above an earthquake's focus.

Different types of earthquakes

Seismic waves

Three types of seismic waves exist.

1 **Primary (P):** fast travelling, low frequency, vibrate in the direction they travel in
2 **Secondary (S):** half the speed of P, high frequency, vibrate at right angles to direction of travel
3 **Surface (L):** slowest of the three, low frequency, some have a rolling movement that moves the surface vertically while others move the ground at right angles to direction of movement

These three types of waves indicate why seismic forces can be so hazardous, as they move the ground in different ways.

Depth of focus

Shallow-focus earthquakes

These common earthquakes occur at the surface to about 70 km depth in cold, brittle rocks. Many release only low levels of energy, but they can be high-energy events.

Deep-focus earthquakes

These earthquakes occur at a depth of 70–700 km. It is difficult to obtain data, so they are poorly understood, but water may play a significant role in releasing energy in these powerful earthquakes.

Assessing earthquake energy

The Richter scale has been in use since the 1930s. Richter only gives an indication about the damage an earthquake might result in via its measurement of energy released. The Modified Mercalli Intensity (MMI) scale measures both earthquake intensity and impact but is a qualitative assessment based upon observation and description.

Increasingly, the moment magnitude scale (Mw) takes precedence. It measures the energy released more accurately than Richter, however, it is not useful when dealing with low-energy events. In any case, these events only register on seismographs and people tend not to feel them.

The effects of earthquakes on landforms and landscapes

Over geological time and on a large scale, earthquakes are part of the processes forming fold mountain chains, such as the Alps. When the African plate moved slowly north to meet the Eurasian plate it put enormous stress on the rocks and caused significant faulting. This continues today, with earthquakes an ongoing threat to certain countries.

East Africa's spectacular Rift Valley indicates the effects of earthquakes on landscapes. The inward-facing escarpments are weathered and eroded. Over time they can blend into the landscape and even disappear altogether as sediment builds up.

Earthquakes generate distinctive hazards

On average, there are 100 earthquakes per year globally that release enough energy to significantly impact people. As with volcanic eruptions, an individual earthquake does not pose the complete range of possible hazards.

- **Ground shaking and ground displacement:** this is the vertical and horizontal movement of the ground. How severe this is depends on:
  - earthquake magnitude
  - distance from the epicentre
  - local geology

Richter scale Measures wave amplitude to assess the energy released when rocks fracture. It uses a logarithmic scale from 0.0 to just over 9.0, the largest earthquake yet recorded.

Modified Mercalli Intensity (MMI) scale Quantifies what was felt by people and the type and scale of damage to buildings on a scale from I (not felt) to XII (total destruction).

moment magnitude scale (Mw) Measures the amount of physical movement of the ground to assess the intensity of an earthquake.

Seismograph Instrument used to detect and record seismic waves.

Locations close to the epicentre of a high-magnitude earthquake receive the most serious impacts, especially if the local geology is made up of unconsolidated sediment (for example, in Mexico City (1985) and Kobe, Japan (1995)). Generally, it is horizontal movement that is the greatest threat to buildings — once a building starts to sway considerably, it can collapse and crash into neighbouring structures. Movement along a fault line can fracture pipelines and sewers, and break rigid structures such as railway tracks and roads.

Ground movement, such as rivers or stream being diverted, can disrupt natural drainage. The movement of underground water in aquifers can be altered. These changes have implications for water supply and irrigation.

- **Liquefaction:** if the surface material at a location is made up of unconsolidated sediments, such as fine-grained sands, alluvium or even landfill, and has a high water content, earthquake vibrations can cause the material to behave like a liquid. Consequently, the material's strength is greatly weakened, resulting in riverbanks collapsing and structures tilting and sinking as their foundations give way.

- **Landslides and avalanches:** slopes give way if they are shaken too much. Steep slopes in mountainous regions, such as Nepal, are notoriously unstable — Nepal is in a very active seismic zone but there has also been much deforestation, which removes the binding effect tree roots have on the slopes. Landslides can block rivers in valley bottoms, forming a dam. As the water builds up upstream of the landslide, flooding from the temporary lake can threaten the valley downstream. Landslides following an earthquake frequently disrupt transport routes in regions such as the Andes. Upland valleys are often used for reservoirs. An earthquake can threaten the stability of a dam either by weakening the dam through shaking or by causing a landslide to crash into the reservoir. This can send a high-energy wave over the top of the dam, causing it to weaken and possibly fail, flooding the valley below.

- **Tsunamis:** underwater earthquakes can cause the sea bed to rise. The water immediately above is displaced and powerful waves spread out. As these waves have a low height (< 1 m) and very long wavelength (up to 200 km), they can race over the ocean surface undetected by ships. However, once they enter shallow water wave height increases rapidly until the wave breaks, releasing a wall of water that crashes onto shore. The wave can exceed 25 m in height and carry hundreds of tonnes of water per metre. The resulting devastation can be extreme and cause extensive flooding, both along the coast and inland along river valleys.

Underwater landslides can also cause tsunami waves. When a large volume of rock is shaken free and slides downslope it drags large volumes of water with it. A wave results from the collision of water.

Alluvium The general term given to material, usually clays, silts and fine-grained sands, deposited by a river when it overflows its banks and spreads across its floodplain.

Knowledge check 31

What is liquefaction?

What are the implications of living in tectonically active locations?

The potential hazards of volcanic eruptions and earthquakes have not stopped millions of people living in locations that are known to be tectonically active. Neither earthquakes nor volcanic eruptions can be prevented, so societies try to develop resilience to cope with the impacts.

Impacts people experience as a result of volcanic eruptions

The level of risk people face in regards to volcanic eruptions is directly related to the type of volcano and volcanic eruption.

Active, dormant and extinct volcanoes

The likelihood of a volcano erupting is indicated by the terms active, dormant and extinct.

- **Active:** erupted in past 10,000 years
- **Dormant:** has not erupted in past 10,000 years but is expected to erupt again some time in the future
- **Extinct:** not expected to erupt again

However, it is misleading to believe that volcanoes can be placed into one of these categories with certainty. Records of past eruptions over such long timescales are fragmentary and not always reliable nor accurate. For example, uncertainty exists where there is evidence of underground magma movement but no eruption. Such questions add to the risks from volcanic activity.

Case study

For this part of the specification you are required to have case studies of **two countries** at contrasting stages of economic development. Examples include: ACs (e.g. Japan, Iceland, Italy, the USA), EDCs (e.g. Costa Rica, Guatemala, Indonesia, Mexico, Montserrat) and LIDCs (e.g. Cameroon, Cape Verde, Democratic Republic of Congo, Ethiopia). Each case study must illustrate:
- reasons why people choose to live in tectonically active locations
- the impacts people experience as a result of **volcanic eruptions**
- economic, environmental and political impacts on the country

Impacts people experience as a result of earthquake activity

The level of risk people face in regards to earthquake activity is directly related to the type of earthquake and where it strikes.

Case study

For this part of the specification you are required to have case studies of **two countries** at contrasting stages of economic development. Examples include: ACs (e.g. Japan, Italy, the USA), EDCs (e.g. China, Iran, Mexico, Pakistan) and LIDCs (e.g. Haiti, Nepal). Each case study must illustrate:
- reasons why people choose to live in tectonically active locations
- the impacts people experience as a result of **earthquake activity**
- economic, environmental and political impacts on the country

What measures are available to help people cope with living in tectonically active locations?

With global population heading towards 10 billion by 2050, and as more people live in towns and cities, greater numbers of people and property are at risk from earthquakes and volcanoes. Many seismic zones are areas with high population densities.

As the risk increases, the ability to reduce risk also increases. Mitigation and resilience are key factors in helping people cope with tectonic hazards. Resilience varies greatly between and within communities.

Resilience The ability of countries, communities and households to resist, absorb and recover from shock and stress.

People's exposure to risks and their ability to cope with tectonic hazards change over time

Exposure and vulnerability to tectonic hazards

Geophysical events such as volcanic eruptions and earthquakes become hazards when they pose a risk to people. How exposed and vulnerable a particular community or household is to a hazard can be indicated by the **disaster risk equation**:

$$\text{Risk } (R) = \frac{\text{frequency or magnitude of hazard } (H) \times \text{level of vulnerability } (V)}{\text{Capacity of population to cope and adapt } (C)}$$

or:

$$R = \frac{H \times V}{C}$$

A disaster is an event when a hazard exceeds the capacity of a country, community or household to cope with the effects of that hazard.

Physical exposure to earthquakes and volcanoes depends on:

- frequency of earthquake and/or volcanic eruption
- magnitude of earthquake and/or volcanic eruption
- types of hazards generated by the event in a particular location
- number of people living in an area prone to tectonic events

The relationship between magnitude (energy released) of a tectonic event and its impact is influenced to some extent by how often the event occurs and the time interval between such events (see Figure 11).

High-energy events occur less often and generally occur less frequently (recurrence interval). Even those with the greatest resilience can be seriously affected by high-energy events — a major earthquake striking southern California will have devastating impacts across the region, despite the wealth and preparation of the people living there. However, the most vulnerable are those who experience serious socioeconomic impacts from relatively small physical changes.

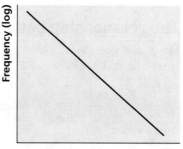

Magnitude of event

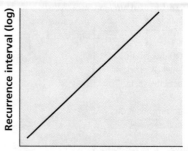

Magnitude of event

Figure 11 The relationship between magnitude, frequency and recurrence interval of tectonic hazards

How and why have risks changed over time?

The Emergency Events Database (EM-DAT), maintained by the Belgium-based Centre for Research on the Epidemiology of Disasters (CRED), has existed since 1988. Its data show an increase in geophysical **disasters** (earthquakes, volcanoes and mass movements) through time (it is important to note that not all mass movements result from tectonic events).

Given the timescales over which tectonic forces operate, we must carefully interpret graphs of past earthquake and volcanic events. Over the past 50 years, there has been an increase in the number of tectonic disasters, currently running at about 30/year. However, this is not evidence that tectonic activity is rising, rather that more people are living in seismic zones with the consequent increase in building construction and economic activity in the same locations.

In terms of fatalities, numbers affected and economic cost, the pattern is of a few very high-energy events causing the most serious disasters. The 2004 Indian Ocean Boxing Day tsunami following a 9.2 Mw underwater earthquake and the 2010 Haiti earthquake (7.1 Mw) resulted in some 230,000–250,000 and 220,000 fatalities respectively. The 2011 Tōhoku earthquake (9.0 Mw) off the northeast coast of Japan is estimated to have cost some £181 billion in damages.

There is a clear contrast in risks posed by earthquakes and volcanoes, with the former representing a much higher threat, especially as regards injury and death. Volcanic eruptions can have serious economic impacts. For example, the ejection of vast quantities of ash can severely disrupt transport, especially air traffic.

> CRED defines a **disaster** as > 10 people killed and/or > 100 people affected.

> **Knowledge check 32**
>
> Why has there been an increase in the number of disasters arising from tectonic hazards?

The relationship between disaster and response

One way of generalising how a disaster affects people is via the disaster-response curve (the Park model — see Figure 12).

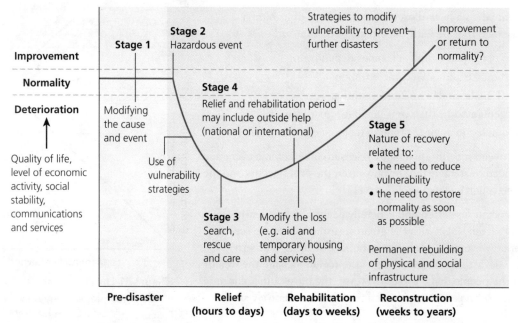

Figure 12 The disaster-response curve (source: Park (1991))

The particular shape of the curve varies according to a number of factors.

- Physical factors:
  - speed of onset of tectonic event
  - magnitude of event
  - length of time the event lasts
- Human factors:
  - quality and quantity of monitoring along a seismic zone
  - degree of preparation
  - quality and quantity of relief

Exam tip

Durham University's Institute of Hazard, Risk and Resilience website is a valuable source of up-to-date information when investigating hazard impacts: www.dur.ac.uk/ihrr/.

Three groups of strategies summarise approaches to managing tectonic hazards (see also Table 18):

1 modify the event

2 modify people's vulnerability

3 modify people's losses

Table 18 Some strategies for managing tectonic hazards

Modify the event	Modify people's vulnerability	Modify people's losses
Not possible for the vast majority of volcanic eruptions. However, the following have been tried with some success: - lava-diversion channels - spraying lava to cool it so that it solidifies - slowing lava flows by dropping concrete blocks Earthquakes: - Nothing can be done to modify an earthquake event.	- **Education:** recognise signs of possible eruption, what to do when an eruption occurs, e.g. evacuation routes, drills to practise what to do when a tectonic event strikes, e.g. in an earthquake, get to open space away from buildings or shelter under a table in a doorway - **Community preparedness:** e.g. building tsunami shelters and walls, strengthening public buildings, e.g. hospitals, fire stations, schools - **Prediction and warning:** increasing use of technology to monitor particularly active locations, e.g. individual volcanoes - **Hazard-resistant building design:** e.g. cross-bracing of buildings to support them during an earthquake, steep sloping roofs to prevent ash build-up - **Hazard mapping:** e.g. predicted lahar routes, ground likely to liquefy in an earthquake - **Land-use zoning:** to prevent building in locations identified by hazard mapping	- **Emergency aid:** e.g. bottled water, medical supplies, tents, food packs - **Disaster-response teams and equipment:** e.g. helicopters and heavy lifting machinery - **Search and rescue** strategies - Buildings and business **insurance** - **Resources for rebuilding public services:** schools and hospitals, and help for individuals to rebuild homes and businesses

People can do little to modify either a volcanic eruption or an earthquake. The energy involved in such tectonic events is so vast that attention is focused on modifying vulnerability and loss.

Building design and the earthquake risk

Much attention is now given to modifying vulnerability to earthquakes with the use of **aseismic design**. A wide range of techniques is being developed to design buildings and structures that stand up to certain degrees of ground shaking and displacement. Such research and practice can be found across the development continuum. However, cost and the ability of authorities to enforce building regulations both play a key role in determining how effective this measure is in reducing people's vulnerability.

Knowledge check 33

What is aseismic design?

Content Guidance

Case study

For this part of the specification you are required to have case studies highlighting various strategies to manage tectonic hazards. You require **an example each of seismic and volcanic** hazards from **two countries** at contrasting levels of economic development. Examples include: ACs (e.g. Japan, Italy, the USA), EDCs (e.g. China, Indonesia, Mexico) and LIDCs (e.g. Haiti, Nepal, Nicaragua). Each case study must illustrate:

- attempts to mitigate against the event
- attempts to mitigate against vulnerability
- attempts to mitigate against losses

Summary

- There is a variety of evidence for the theories of continental drift and plate tectonics.
- At the different types of plate boundaries there are distinctive processes and features.
- There is a variety of volcanic and earthquake activity, which leads to different landforms and landscapes and distinctive hazards.
- People experience a range of impacts as a result of volcanic eruptions and earthquake activity.

- The exposure of people to risks and their abilities to cope with tectonic hazards changes over time (the disaster risk equation).
- There is a relationship between disaster and response (the Park model).
- There are various strategies to manage hazards from volcanic and earthquake activity.

Questions & Answers

Assessment overview

For **A-level**, Paper 3: Geographical debates is 2 hours 30 minutes long, carrying 108 marks. This paper makes up 36% of the A-level qualification.

For **AS level**, Paper 2: Geographical debates is 1 hour 30 minutes long, carrying 68 marks. This paper makes up 45% of the AS qualification.

At AS and A-level, Geographical debates covers the same five options: Climate change, Disease dilemmas, Exploring oceans, Future of food and Hazardous Earth. At both levels the paper has three sections:

■ Section A: short-answer and medium-length questions on all options
■ Section B: synoptic questions on all options
■ Section C: extended response (essay) questions on all options

If you are taking the **A-level**, you will have to answer questions from each section on **two** options of your choice.

At **AS**, you also answer questions from each section but on just **one** option of your choice.

A-level

Questions in Section A consist of two parts: part (a) requires you to identify limitations with the data presented in a resource, for 3 marks. The resource could be quantitative data such as statistical maps, diagrams and tables, or qualitative information in the form of photographs or brief written extracts. Part (b) assesses your knowledge and understanding of some aspect of the option, for 6 marks. It is recommended that you allocate about 25 minutes to your two chosen questions (four parts).

Section B has just one question for each option, which carries 12 marks. This synoptic section assesses your ability to draw together knowledge and understanding from across the whole course, such as a landscape systems, Earth's life support systems, changing spaces, making places or global connections. Giving examples to support your points will often help to make your answers more convincing. It is recommended that you allocate about 35 minutes to your two chosen questions.

Section C has two questions for each option, from which you choose one. These extended response questions (denoted *) are worth 33 marks each and they focus on your ability to analyse and evaluate. It is recommended that you allocate about 45 minutes to each of your two chosen questions.

AS

Section A consists of one question per option. The question will be a combination of short-answer questions, with parts typically worth 4 or 6 marks but with the final part, part (d), worth 12 marks. Part c (i) is based on some numerical data and will require you to carry out a calculation, such as mean, median, mode, range or interquartile range. This will carry 4 marks, while 6 marks are given for an analysis of the data in part c (ii). It is recommended that you allocate about 35 minutes to this section.

Section B consists of one question per option, divided into two parts. Part (a) is based on a resource such as quantitative data, for example statistical maps, diagrams and tables, or qualitative information in the form of photographs or brief written extracts. Both parts carry 8 marks and assess your ability to draw together knowledge and understanding from across the whole course, such as either a landscape system, changing spaces or making places. Giving examples to support your points will often help to make your answers more convincing. It is recommended that you allocate about 20 minutes to this section.

Section C has two questions in each option, from which you choose one. This extended response question (denoted *) is worth 20 marks and focuses on your ability to analyse and evaluate. It is recommended that you allocate about 35 minutes to this question.

At both AS and A-level, the quality of your extended response will be assessed so, as well as needing to show comprehensive knowledge and understanding of the topic, you should use full sentences, spell and punctuate correctly, and make appropriate use of technical terminology.

Geographical skills will be assessed within this component — these are identified in the specification.

About this section

The questions below are typical of the style and structure that you can expect to see in the A-level paper. Each question is followed by examiner comments, which offer some guidance on question interpretation. Student responses are provided, with detailed examiner comments on each answer, to indicate the strengths and weaknesses of the answer and the number of marks that would be awarded. A final summary comment is also provided, giving the total mark and grade standard.

Section A

Climate change

Question 1

Extract 1

There is actually very little carbon dioxide in the atmosphere (0.038% of total gases). It has, however, increased rapidly over the past 200 or so years. The past 20 years have seen annual increases of about 30 ppm, most of which is due to human activities.

Other greenhouse gases are significant in causing global warming and most of these are naturally present. Human activities also influence their concentrations and therefore global warming.

(a) Identify three limitations of Extract 1 in identifying the causes of climate change. [3 marks]

(b) Explain how geoengineering may reduce the impact of global warming. [6 marks]

Student answer

(a) A phrase such as 'increased rapidly' is not really useful, as it doesn't tell you anything about the levels of carbon dioxide in the atmosphere. The annual increases are also not useful, as the text doesn't say what the actual level of carbon dioxide is ▪a. Simply saying 'other greenhouse gases are significant' is not useful, as it doesn't say what they are ▪b. The text says that these are 'naturally present' but that human activity 'also influences their concentrations', however you don't know if these gases are increasing or if it is due to natural causes or humans ▪c.

ℯ 3/3 marks awarded. It's good that the student quotes the text more than once, as this makes it clear to the examiner that s/he is making direct use of the resource. ▪a The limitations of the statistics, ▪b the vagueness of some statements and ▪c the comment about natural or human causes are appropriate limitations to highlight.

(b) Geoengineering is when scientists use technology to stop global warming. There are two types of geoengineering, one way is to stop so much solar energy from coming into the atmosphere and the other is to reduce the amount of greenhouse gases such as carbon dioxide being produced by humans ▪a. Stopping solar energy is very difficult and may involve putting giant mirrors out in space to reflect the sun's rays. This will be very expensive and will need lots of countries to cooperate in paying for it ▪c. Another way is to inject billions of small particles into the atmosphere, which will intercept the sun's rays before they reach the surface and scatter the rays. Much of the sun's energy will be sent back into space and so reduce global warming ▪b.

Some people have suggested that we should spread iron particles in the oceans, called fertilisation. This will mean that the oceans will have more nutrients and will allow organisms like phytoplankton to grow. These microscopic organisms absorb carbon dioxide because they photosynthesise ⓑ. There is also an idea of artificial trees made of a plastic that can absorb carbon dioxide, which is then stored underground.

ⓔ **6/6 marks awarded = Level 3.** ⓐ The answer deals directly with the topic of geoengineering, starting by defining what it is. ⓑ The student then shows that their knowledge and understanding of the topic is thorough and well-developed by referring to several geoengineering techniques. The student has kept a sharp focus on the command word 'explain' and not drifted off into analysis of the feasibility of geoengineering, ⓒ apart from in one sentence.

Disease dilemmas

Question 2

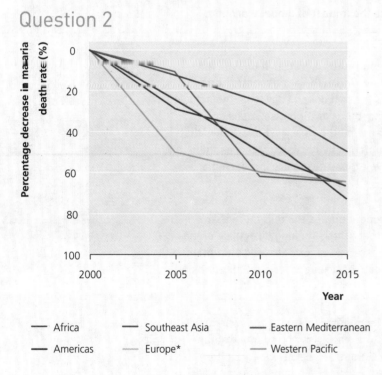

Figure 1 Percentage decrease in malaria death rate by WHO region, 2000–2015 (source: WHO *Global Malaria Report*, 2015)

Legend:
— Africa — Southeast Asia — Eastern Mediterranean
— Americas — Europe* — Western Pacific

* There are no recorded deaths among indigenous cases in the WHO European region for the years shown.

Study Figure 1, percentage decrease in malaria death rate by WHO region, 2000–2015.

(a) Identify **three** limitations of Figure 1 for showing decline in malaria death rates. [3 marks]

(b) Explain the influence of physical factors on the global distribution of malaria. [6 marks]

Student answer

(a) Malaria decline is shown in Figure 1 by percentage change over time. The graph does not provide absolute numbers of deaths or death rates ⓐ. The WHO regions are extensive areas — there is no indication of variation in malaria death rates within them ⓐ. The long 5-year periods used for analysis of the data give no evidence of interim fluctuation ⓐ.

ⓔ **3/3 marks awarded.** ⓐ The response identifies three limitations of the data evidence.

(b) Incidence of malaria is greatest where there are natural breeding habitats for mosquitoes. These include tropical and subtropical areas of sub-Saharan Africa, south Asia, southeast Asia and Latin America ⓐ. In these localities transmission of malaria by the female Anopheles mosquito vector is greatest where temperature and rainfall create warm, moist conditions ⓑ. These conditions are found all year round near the equator ⓑ.

Further away from the equator, malaria transmission occurs at the end of the rainy season ⓐ. This is when mosquitoes can lay their larvae in any stagnant water after flood or monsoon ⓑ.

Risk of malaria is lower where there is high altitude, aridity or a cold season ⓐ. In these conditions the Plasmodium falciparum parasite cannot complete its growth cycle in mosquitoes and is not transmitted ⓑ.

ⓔ **5/6 marks awarded = Level 3.** ⓐ This response demonstrates thorough knowledge and understanding of factors. ⓑ Explanation is well developed, although not supported by climatic statistics.

Exploring oceans

Question 3

Table 1 Seabird population trends for selected species around the UK coastline

Species	Population change 1998–2015 (%)
Arctic tern	+17
Black-headed gull	+38
Great cormorant	−8
Northern fulmar	−31
Razorbill	+32

(a) Identify three limitations of Table 1 in identifying how healthy the marine ecosystem is around the UK. [3 marks]

(b) Explain the pattern of principal shipping routes across the oceans. [6 marks]

Student answer

(a) The figures are only for a few species of seabird and it doesn't tell you about all of the marine ecosystem [a]. The table is also giving different figures for the species, some are increasing such as razorbill but some are declining, e.g. northern fulmar [b].

e **1/3 marks awarded.** [a] The response gives a suitable limitation in the first sentence. [b] However, the second point about variations in the figures in terms of growth or decline is not a limitation of the data, rather representative of the complexities in the real world. No third limitation is offered.

(b) Oceans are crossed by many shipping routes. One of the main routes is between Europe and North America [a]. This is because these regions are very wealthy and there is a vast amount of trade between them. The GNI per person in the EU is about $38,000 while the USA is $57,500. The UK is $40,600, so people can buy goods from each other that need to be shipped across the Atlantic [b]. There is also trade between Europe and Asia, and this is where physical geography affects shipping routes [b]. One way to travel is right round the Cape of Good Hope but when the Suez Canal was built, it meant a shorter distance for boats. There is also a lot of shipping going across the Pacific and this is affected by the Panama Canal, as this means boats do not have the very long journey around Cape Horn [a]. Another important shipping route goes from Europe to Brazil [a]. This is because a lot of trade is between them and also the USA. Brazil exports vast amounts of minerals, such as iron ore and bauxite, as well as primary goods, such as agricultural products [b].

e **6/6 marks awarded = top Level 3.** This is a thorough answer focused on the question. Substantial knowledge is given of both [a] the actual routes and the reasons they exist. [b] The use of figures is supportive and relevant.

Future of food

Question 4

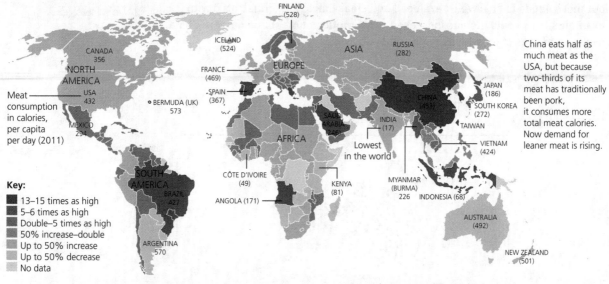

Figure 2 Change in meat consumption (1961–2011)

(a) Identify three limitations of Figure 2 showing change in global meat consumption from 1961 to 2011.

[3 marks]

(b) Explain the impact of globalisation on dietary patterns.

[6 marks]

Student answer

(a) While the graduated shading is easy to understand, the intervals in the key are confusing – the first category is 13–15 times, but the second is 5–6 times, an interval of two then one ⓐ. The wording then changes from 'times' to percentages ⓐ. The shades for up to 50% decrease and no data are very similar and difficult to interpret ⓐ.

ⓔ 3/3 marks awarded. ⓐ The response gives a range of suitable limitations, expressed with clarity. Three covered so 3 marks.

(b) Globalisation has led to increased wealth and affluence in many parts of the world due to increased inward investments, job creation by TNCs and new trade opportunities ⓑ. As countries develop and affluence increases people have more money to spend on food, as they have more disposable income. So they are better fed ⓐ. More food products are traded as communication and transport improves and trade deals such as multilateral and bilateral agreements are signed.

🅔 **2/6 marks awarded = top Level 1.** This is a basic answer regarding the focus of the question on the impact on diet. 🅐 Basically, the only point offered is a very vague people are 'better fed'. What exactly does this mean in terms of quality and quantity of food? 🅑 The knowledge on the process of globalisation is better, but there needs to be a well-developed and clear link to diet, illustrated with examples. This could include increased consumption of meat and dairy in China, or wider choices of food types due to global travel.

Hazardous Earth

Question 5

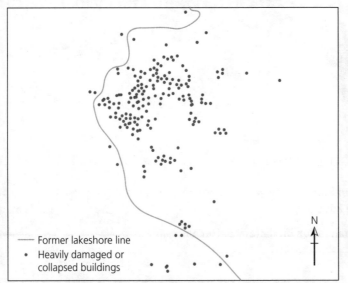

Figure 3 Location of heavily damaged or collapsed buildings in central Mexico City following the 1985 earthquake

(a) Identify three limitations of Figure 3 in identifying the impact of an earthquake. [3 marks]

(b) Explain the variety of earthquake activity. [6 marks]

> **Student answer**
>
> **(a)** One limitation of the map is that it only shows the heavily damaged and collapsed buildings. It does not show where more minor damage happened 🅑. The map does not have a scale so it is impossible to see what sort of area the damage covered 🅐. One other limitation is that other impacts such as deaths or injuries are not given, so you don't know all the impacts 🅑.

🅔 **3/3 marks awarded.** 🅐 The response focuses on the resource and correctly identifies a technical cartographic limitation in the lack of a horizontal scale. 🅑 The other two limitations are focused on data limitations when assessing the impacts of the earthquake.

(b) When earthquakes happen, huge amounts of energy are given off. This is because friction builds up in the Earth's crust and then there is the sudden slipping of plates, which gives off the energy. When plates converge one is usually subducted below the other. This causes the subducted plate to melt and break up, and this releases earthquake energy **a** along the Benioff zone **b**. Earthquakes also happen when plates pull apart **a**, although they are often not as strong as at a destructive boundary. Iceland, for example, does have earthquakes but most of these are low-energy ones. The Himalayas have many earthquakes (Nepal) and this is because it is a collision zone. Two plates have crashed into each other and earthquakes happen as the rocks grind against each other **a**. California has many earthquakes, as it is a conservative plate boundary **a**.

e **4/6 marks awarded = Level 2.** While this response has relevant material, the knowledge and understanding it contains is reasonable rather than thorough. **a** There is mention of a variety of earthquakes with the reference to those occurring at destructive and constructive plate boundaries. However, the answer is not displaying a real authority over the material as it tends to be rather vague — there is no detail as to what type of plate movement exists in California, for example. There is a lack of technical detail such as of shallow and deep focus earthquakes, **b** although the correct use of the term Benioff zone is encouraging.

■ Section B

Climate change

Question 6

Assess how climate change may bring about changes to the water cycle in tropical rainforests.

[12 marks]

> **Student answer**
>
> Tropical rainforests are one of the most important ecosystems on the planet. They cover the land around the equator and extend up towards the tropics. The Amazon is an example. The water cycle in the rainforest is very active as there is so much water and high levels of energy. The energy comes from the sun, which is very intense all year and means that the temperature is high, usually above 25°C a. The high temperatures mean that there is a lot of evaporation every day from any water on the surface. This means there is a huge amount of water vapour going into the atmosphere above the rainforest, which means that clouds can form and more rainfall happens a. The trees also give off water vapour.
>
> Climate change is happening all over the world. Because human activities are producing more greenhouse gases (carbon dioxide and methane), more of the sun's energy is being kept in the atmosphere and so global warming is taking place. This means that more evaporation can happen and this affects the water cycle. This might mean that there is more rainfall but because the air is warmer, it may not rain as much b.
>
> In the rainforest, global warming might result in it becoming less wet as the pattern of rainfall changes c. It could also be the result of deforestation, as this alters the water cycle and means that less rain occurs. If this happens then more of the rainforest will become grassland. Deforestation can be because humans are clearing the forest but it could also happen naturally, as there is less rainfall and the trees die d.

e **7/12 marks awarded = top Level 3.** **a** The response starts promisingly with the focus on the water cycle in the tropical rainforest and some encouraging knowledge and understanding shown. **b** The material on climate change has potential but lacks detail, so that the student's knowledge and understanding is only reasonable. The crucial area of making synoptic links between climate change and the rainforest's water cycle is present but not as clear as it might be '… as the pattern of rainfall changes'. **c** Not writing separate paragraphs about the two topics, climate change and rainforest water cycle, is more likely to lead to a response linking the topics more successfully. **d** There is some analysis, as the command word 'assess' indicates there should be, regarding human influence on the rainforest's water cycle, but this is not substantial.

Disease dilemmas
Question 7

'Poverty is the main factor in the spread of infectious disease.' How far do you agree with this statement?

[12 marks]

Student answer

A wide range of factors affect patterns of infectious diseases — of these, poverty is highly significant. The relatively poor governments of LIDCs and some EDCs may be unable to provide sufficient healthcare resources for their populations [a]. Numbers of hospitals, local medical facilities, trained health workers and medicines such as vaccines may be in short supply simply because of costs [a]. As a result, populations most affected by the spread of viruses are rural communities which have limited accessibility, for example, the rural poor in sub-Saharan Africa and parts of southeast Asia and those living in high-density urban slums [a]. There may be limited access to clean drinking water leading to the spread of water-borne infections such as bilharzia [a].

Poor households, especially families living in LIDC urban areas, are at high risk of infectious disease. Viruses spread rapidly where people live in close contact in overcrowded conditions or where there is poor sanitation. Inadequate diet leading to undernutrition and malnutrition in poor families weakens immunity and also contributes to rapid spread of disease following an outbreak [a]. For example, influenza, diarrhoeal disease, malaria and TB are the major causes of death from infectious disease in Malawi.

The range of factors other than poverty that influence patterns of infectious disease can be illustrated by the Ebola outbreak in west Africa in 2013. The initial inability of medical staff to recognise the disease was significant — some health workers and doctors became infected and their subsequent attempts to assist in outlying settlements spread the virus further [b]. Daily population movements to local markets in the neighbouring countries of Guinea, Liberia and Sierra Leone also spread the virus rapidly [b]. Traditional cultural customs of washing, kissing and touching at funerals caused further transmission of the disease [b]. Years of civil war in Sierra Leone destroyed infrastructure including medical resources [b]. Also many local communities were suspicious of outside intervention, which added to the delay in specialist help from ACs [b].

In Ethiopia, malaria infection increases when large numbers of people move between the highlands and lowlands at harvest time. The migrant workers are at greatest risk of malaria transmission in the lowlands, especially since harvesting continues after sunset when the mosquitoes are most active [b]. Other factors include the effects of climate change on vector habitats, which may lead to emerging infectious diseases. Examples include the northward spread of West Nile virus and Lyme disease in North America, and the southward spread of sleeping sickness into southern Africa as temperatures increase [b]. Poverty has a significant impact on disease patterns, especially in LIDCs, which are at an early stage of the epidemiological transition, but this is just one of many social, economic, political and environmental factors that contribute to the spread of infectious disease.

ⓔ 12/12 marks awarded = top Level 4. **ⓐ** There is comprehensive knowledge and understanding of ways in which poverty contributes to spread of infectious disease. **ⓑ** There are well-developed attempts to make synoptic links between content from different parts of the course.

Exploring oceans

Question 8

Examine how the players (stakeholders) involved in an oil spill influence economic change in places affected.

[12 marks]

> **Student answer**
>
> When an oil spill happens there are many players involved. If the oil spill comes from a tanker, the owners **ⓐ** are involved in the clean-up operation. The same happens if the oil comes from a drilling rig, for example Deepwater Horizon in the Gulf of Mexico. BP has had to pay millions of dollars for the clean-up as well as compensation to local people such as fishermen. The fishermen **ⓐ** were unable to trade, as where they fish was polluted by oil. This meant they lost money and some became unemployed **ⓑ**.
>
> Another group affected by this oil spill was the tourist industry **ⓐ**. As there was so much oil washed up on the beaches, visitors didn't come and so hotel staff and restaurant workers became unemployed **ⓑ**. However, with the massive clean-up operation and the compensation BP paid, the beaches are being cleaned so that tourists can return and the economy start to grow.
>
> The government **ⓐ** is also involved in oil spills as it helps to organise operations. In the USA, NOAA coordinates the restoration programmes as well as the state authorities. It is trying to make sure that the places affected recover as quickly as possible so that economic change can happen. For example, it has been working on cleaning up saltmarshes and the areas of sargassum, as this is where many fish breed. This will then allow the fishing industry to recover, as well as bringing tourists who want to fish.
>
> As Deepwater Horizon blew up and sank, oil production stopped. This means that there can be negative economic change, as people employed in the oil industry may lose their jobs **ⓑ** . However, it might be that jobs are available in other places.

ⓔ 9/12 marks awarded = Level 3. **ⓐ** The response is encouraging as it is focused on the question throughout, dealing with relevant players **ⓑ** and economic change. The use of a case study is helpful, particularly as it does not follow a 'write all I can remember' approach. Rather the student tries to manipulate the material so that the link between players and economic change in places is established. However, the discussion would be more convincing with the addition of some more detail about the example, such as names of some of the places/states affected.

Future of food

Question 9

Examine how the carbon cycle is affected by strategies to improve food security. [12 marks]

> **Student answer**
>
> Strategies to improve food security include the MERET programme. A joint venture between the WFP and the Ethiopian government, this project aims to reclaim degraded land and increase food production [a].
>
> ■ Terracing of hillsides to prevent soil erosion: when soil is left bare and eroded there is not sufficient soil and nutrients (which are in the upper soil layer) to grow crops. Soil is a carbon store and if it is exposed and eroded by wind and/or water, the carbon store is reduced. Less soil erosion will help maintain carbon stores and improve fertility for farming [b].
>
> ■ Installing 'cooking stoves', which burn three times less wood, reduces deforestation. This is a good thing as trees absorb carbon dioxide from the atmosphere and sequester [c] it for hundreds of years. Deforestation releases this carbon back to the atmosphere, with negative consequences for climate change and global warming [b].
>
> Pressure to increase productivity from farmland in many countries including ACs impacts the carbon cycle, as successive harvests mean a loss of carbon, as very little carbon is returned to the soil through organic matter [b].
>
> There are also low levels of carbon exchange through the process of photosynthesis in the 'fast carbon' cycle [c], as growth of crops takes 4–5 months.

ⓔ 7/12 marks awarded = bottom Level 3. [a][b] The response tips into Level 3 through the use of a developed, specified example, which clearly relates the attempts to produce more food (increasing food security) to impacts on the carbon cycle and provides a reasoned account rather than a list of facts. **[c]** There is also good use of appropriate subject-specific terminology in the right context, e.g. 'sequestration' and 'fast carbon cycle'. However, the answer fades somewhat in the second part and becomes generalised. The use of bullet points in the first part of the response is interesting — the bullets help to structure the answer and are followed by developed information, showing that it is not the use of a bullet point that is an issue — it depends on what follows it.

Hazardous Earth

Question 10

Examine how living in a seismically active place influences how people understand that place.

[12 marks]

Student response

Many millions of people live in places where volcanoes and earthquakes happen, such as Japan and Nepal. Sometimes these earthquakes are so powerful that large numbers of people are affected — for example, about 9,000 people were killed in Nepal in 2015. This meant that people living outside of Nepal thought that it was a dangerous place to live. It also meant that tourists or climbers may not visit because of the earthquake risk a.

People living in a place that has lots of seismic activity can prepare for an earthquake. For example, in California they have earthquake drills in schools and so people know what to do b. At home they secure their furniture to the walls and much of the building just has one storey, so if a building collapses it is not as serious as if a whole block of flats fell down. Tall buildings are made as safe as possible by measures such as cross-bracing and rubber foundations to absorb earthquake waves. People make their place as safe as possible b.

People often have a strong emotional attachment to the place where they were born and grew up. They know it is risky because of earthquakes but they prefer to live there rather than somewhere else. In Nepal, it may be that their family has owned a farm for many generations and so they don't want to leave c. California has a very attractive warm, sunny climate. There are also many employment opportunities and so people accept the risk but need to have a job c. Also in California, the mapping of exactly where the faults are is quite advanced and there are places away from the faults that are safer to live in. In LIDCs such as Nepal, many people have no choice but to live where they do as they are so poor. Even if they migrate to a town, they are still at risk from earthquakes as all of Nepal is a seismic zone. However, it may be that they think that they'll be safer there as if an earthquake happens, rescue and aid is more likely to reach them than in a remote village.

e 10/12 marks awarded = Level 4. This answer maintains a focus on the relationship between seismic activity and the ways people can understand place. a It picks up on how and why people perceive places in different ways. The comment about tourists and their perception of Nepal as a dangerous place is relevant. b It also refers to the ways in which residents of a place (Nepal and California) affected by seismic activity perceive where they live. c The point about emotional attachment is well made, as is the need to have work, and both are linked well with the idea that people know about risks from earthquakes and learn to live with these. The final paragraph is very good as it highlights how some groups have little choice in life. With a little more about how perceptions can alter over time, this has the potential to be a top Level 4 answer. For example, as time passes, people's perceptions of risk change (e.g. tourists returning to Nepal).

Section C
Climate change
Question 11

To what extent have international directives such as the Kyoto Protocol reduced the impacts of climate change?

[33 marks]

Student answer

The Kyoto Protocol began in 1997 and was aimed at cutting emissions of carbon, mainly carbon dioxide. It is a legally binding agreement although its first period ended and now it is a voluntary agreement **a**. Other directives also exist, which aim to reduce the effects of climate change.

Climate change is seen as one of the most serious threats because it is probably going to affect so many different things, such as rising sea levels, coral bleaching and food production. Temperatures have been rising for over 100 years and the last 40 years have been rising particularly quickly. The year 2016 was the hottest since records began in the late nineteenth century and the third hottest year in a row **b**.

The causes of climate change are called forcings. Some of these are natural but none of these are responsible for global warming. It is pretty much certain that humans have been changing the balance of GHGs, which is causing more of the sun's energy to be trapped by the greenhouse effect. Burning fossil fuels is a major source of carbon dioxide and methane comes from agriculture such as livestock and rice farming **b**.

e These opening paragraphs immediately take the essay in an appropriate direction. **a** The brief outline of the Kyoto Protocol and **b** the statements about climate change and its causes and impacts are clear and convincing.

When the IPCC started to give its reports, people decided that action needed to be taken to stop global warming. Many countries supported action such as Kyoto, but some important ones didn't **a**. The USA, for example, didn't and now with Donald Trump as president they are against taking action against global warming Russia also didn't sign originally, but has since signed. Russia has now reduced its carbon emissions by just over 50%, which is more than their Kyoto target. Countries like Switzerland and Spain missed their Kyoto target by over 10% **b**.

A major problem with directives like Kyoto is that if the large emitters of GHGs don't sign up then global warming is just going to carry on **a**. China and India are EDCs so they argue that in order to get their living standards to rise then they should not be limited in their emissions. However, China in particular has begun to stop building coal-fired power stations. Although its emissions will still be vast, it is beginning to change its attitudes and this will help reduce the impacts of climate change **b**.

At the Paris Conference in 2015, many countries agreed to limit global warming to 20%. The problem with the Paris Agreement is that countries will set their own voluntary targets, which might not be enough ⓐ. Obama did sign these agreements but Trump wants to take the USA out of this ⓑ. One of the main problems with international agreements is that politics mean that if there is a change of government in a country, the new leaders may not agree with signing ⓐ.

ⓔ ⓐ These next three paragraphs continue the focus on international directives and offer some analysis and evaluation as to how effective they have been. ⓑ It is encouraging that the student has kept up to date with events concerning climate change, such as the Chinese reappraisal of coal-fired power generation and the different direction resulting from political change in the USA.

Carbon dioxide levels are now above 400 ppm, which is higher than for hundreds of thousands of years. One of the problems is that some gases stay in the atmosphere for very long periods and so carry on trapping heat. Directives such as Kyoto and Paris are trying to do something and any reduction in global warming is going to take many decades to happen. However, doing nothing is only going to make the situation much worse and so Kyoto may have a good effect in the future ⓐ.

ⓔ The concluding paragraph's focus on emissions is relevant, as an increasing atmospheric carbon dioxide level is an impact, ⓐ and overall there is good material on some of the issues affecting international directives. However, the response as a whole is thin on impacts such as sea-level rise, food production, disease distribution and geopolitical tensions. More could be made of actions to adapt and mitigate against some of the consequences of climate change.

ⓔ **25/33 marks awarded = close to the A/B boundary.** There is comprehensive knowledge and understanding, so in AO1 Level 4, 7/9 marks. The application of knowledge and understanding to analyse and evaluate is top Level 3, 18/24 marks.

Disease dilemmas

Question 12*

'International organisations have the most important role in the mitigation of communicable diseases.' Discuss.

[33 marks]

Student answer

International organisations such as the WTO, UNICEF, many NGOs, national governments and local communities all have disease mitigation strategies. The aim is to decrease incidence of specific communicable and non-communicable diseases at global, national and local scales. Strategies range from eradicating or eliminating a disease, reducing risk of outbreaks and containing epidemics. The role of international organisations is highly significant, especially for pandemics such as H1N1 or SARS. They also have a critical role in dealing with the impact of epidemics and in the fight against non-communicable diseases such as cancers or diabetes.

The leading UN health organisation is the World Health Organization (WHO), established in 1948. Its importance is illustrated by its vital role in coordinating international public health and national health policies. The WHO provides global leadership in all health matters and encourages partnerships between all other related organisations, at international, national and local scales. Under the UNDP, three original MDGs were directly related to the WHO aims of reducing child mortality, reducing maternal deaths, and halting and reducing the spread of HIV/AIDS. Currently SDG 3 has a central place in disease mitigation by aiming to ensure healthy lives and promoting wellbeing. Almost all of the other 16 SDGs are directly or indirectly related to health [a].

The WHO's work at a global scale includes monitoring diseases throughout the world, mapping global health trends and disease eradication based on research in each of its six regions. The WHO is the best-placed organisation to ensure the poorest countries receive help through dissemination of knowledge about infections such as TB and the way they spread, as well as understanding incidence and possible causes of cancers and type-2 diabetes. The WHO provides technical expertise, advice and education in order to implement its evidence-based guidelines. The WHO also encourages and advises national governments to develop their own national health policies [a].

The WHO has played a leading role in the complete eradication of smallpox by its sustained global strategies. Its current priorities focus on HIV/AIDS, Ebola, malaria, TB and polio eradication, and it aims to mitigate the effects of non-communicable diseases, improve sexual and reproductive health issues, and improve nutrition in partnership with the FAO and the World Bank [a].

The WHO is financed by contributions from member states. This gives the international organisation authority and sustainability, and cements its position as a highly significant organisation in the mitigation of disease [a].

UNICEF, established as a relief organisation for children in 1946, is also an international organisation of huge importance. Its aims include monitoring disease, combating HIV/AIDS, providing technical support and interventions in controlling major global diseases. UNICEF has a major role in responding to emergencies after hazard events by providing fresh water and shelter, medical supplies and treated insect nets in malaria-infested areas [a].

International NGOs have major contributory roles when crises develop quickly. They provide food, shelter, water, medical supplies and medical assistance in areas of humanitarian crisis, such as epidemics, conflict and earthquakes [a]. For example, the British Red Cross had significant input following the 2010 Haiti earthquake in treating victims of the cholera outbreak and building latrines to improve sanitation and hygiene. Oxfam, another international NGO, is very effective in reducing poverty and educating children to protect themselves from disease [c].

Apart from international organisations, state governments have an important role in disease mitigation. The UK government preparedness policy for influenza pandemics based on WHO advice includes guidance on international travel, reduction of personal risk in catching or spreading the virus, and provision of vaccines. It coordinates actions of businesses, local communities and media b.

The Mauritius government effectively eliminated malaria from the island by the early 1970s. Re-emergence of the disease occurred when construction workers inadvertently reintroduced malarial parasites when repairing damage after a tropical cyclone. The resulting epidemic of 1982 forced the government to set up a second elimination campaign. Screening of tourist visitors at the airport, continued surveillance and spraying of breeding sites continues throughout the island c.

The top-down policies of national governments can encounter local resistance but grassroots strategies involving local communities can be equally, if not more, effective in mitigating disease b. For example, the eradication of Guinea worm disease in Ghana is increasingly possible and sustainable as local volunteers are taught how transmission of the disease can be prevented and how to make thorough checks on water filters and water sources c.

International organisations play a significant part in disease mitigation by global monitoring, providing information to national governments, encouraging national planning for epidemics and achieving SDGs. Coordination of these policies and ensuring cooperation of all bodies involved are an essential role of major international organisations. But at other scales the combination of top-down policies of national governments and the input of local communities through their bottom-up strategies are also successful in mitigating many communicable diseases b.

e a This response shows comprehensive knowledge and understanding of the roles of international organisations, NGOs, governments and local communities in mitigating disease. b Application of knowledge and understanding in evaluating the relative importance of these organisations is thorough. c The statement is analysed in the context of place-specific/disease details. There is a well-developed line of reasoning and a logical structure with appropriate introduction and conclusion.

e 27/33 marks awarded = A grade. This response is placed at the top of Level 4 for AO1 and at the top of Level 3 for AO2.

Exploring oceans

Question 13

Discuss the extent to which biological resources within oceans can be managed sustainably.

[33 marks]

Student answer

Nearly 80% of all life on Earth is found in the oceans but because of the fact that so much of the ocean is still to be properly explored, there are possibly large numbers of species to be discovered. There are some biological resources that humans are exploiting and they are not always being managed sustainably.

e A promising opening sentence sets the essay off appropriately. The second sentence of the introduction, however, is suggesting a conclusion, which is not appropriate here.

One resource that is being exploited is krill. These are small crustacea that are part of the ecosystem in the Southern Ocean around Antarctica. They feed on phytoplankton and then are eaten by creatures such as squid, penguins, whales and birds. They live in vast swarms and their total body mass **c** is bigger than all the humans **a**. In the 1970s, humans started fishing for krill. Krill is used in several things such as oil and paste for humans and animal feed. When krill was first being exploited, many people thought that it would follow a boom-and-bust pattern, as so much would be caught that the population would crash. At one point nearly half a million tonnes was caught in a year, but then it fell right back down, although in the twenty-first century more and more krill is being taken **b**.

e **a** This encouraging paragraph contains effective details about an example of a biological resource, krill. **b** It also starts to address the question as regards sustainable management, but a recent figure for the krill catch would be helpful. **c** One point about vocabulary is that the student has used 'body mass' rather than biomass — not a major concern, but it is better to use the right technical terms.

Sustainable means using a resource in a way that meets the needs of the present without taking away the resource for the use of future generations **a**. When krill fishing was growing, countries such as Japan, China and Norway set up an organisation to manage resources around Antarctica. One good thing that it does is to look at the whole ecosystem, as krill is very important to the ecosystem. This is a good approach to sustainable management as it looks at what effects might happen if very large amounts of krill were allowed to be exploited **b**. However, not everyone agrees, as the commercial players want to carry on fishing for krill. The organisation is setting a total allowable catch (TAC) for krill for each year **b**. One problem is trying to supervise the boats catching krill as the area they fish in is so large. Another problem is that krill fishing is concentrated in just a few areas, such as by the Antarctic Peninsula **b**. It would be better if the fishing was spread out, but this is uneconomical at present.

ⓔ **ⓐ** This paragraph explicitly brings in the concept of sustainability using the widely used definition from the Bruntland Commission, which is fine. When using concepts, it is always encouraging to acknowledge that there is often disagreement as to what the concept actually means in practice — different people have contrasting and conflicting views of sustainability. The mention of a management organisation is good (try to learn names), as is an evaluation of its operation. **ⓑ** These few sentences directly answer the question.

> Another resource being exploited is coral reefs. These are found in tropical oceans, as they need high water temperatures **ⓐ**. Tourists come to see and swim in the coral to see all the organisms that live on a reef. However, in some areas tourists have damaged the coral by breaking bits off, sometimes deliberately, and as coral is slow growing this is not sustainable. In some places **ⓑ** they have introduced zoning to allow some tourism in parts of the reef, but ban it in areas kept for research. Coral is also important for fish to breed, which affects food supplies. Also, global warming is affecting coral by bleaching it.

ⓔ **ⓐ** Introducing another resource in a different location is good. However, the paragraph lacks authority due to the absence of detail, such as place names **ⓑ**, and no evaluation of management, such as zoning.

> Many biological resources are under threat, such as fish, and different types of management are used to manage them sustainably. However, with growing populations it is going to be difficult to keep the management sustainable in the future.

ⓔ It is important to finish these essays with a conclusion, however brief. The final sentence is very interesting and the point it raises deserves to be discussed in full earlier in the essay.

ⓔ **20/33 marks awarded = equivalent to C grade.** This essay has scored 5/9 (Level 3 in AO1) for knowledge and understanding and 15/24 (Level 3 in AO2) for application of knowledge and understanding to interpret, analyse and evaluate geographical issues. With more factual detail, such as about reef management, and more analysis of how sustainable that management is, the response would go into Level 4 in AO1 and AO2.

Future of food
Question 14*

To what extent does food security vary within countries?

[33 marks]

Student answer

Food security is about not only physical access to sufficient food but also economic access. The quality of the food as well as the quantity is also important [a]. People may have food to eat, but an unbalanced and unhealthy diet represents a food security issue. The FAO states that people should have access to 'enough safe and nutritious food to meet dietary needs'. This essay will use this definition and will show that countries across the development spectrum face internal variations in food security. A range of examples will show regional differences, e.g. in China and USA, differences within urban areas, e.g. Kenya, and variations between different groups of people within one country, e.g. the USA [b].

e [a] A very useful introduction that defines the key term of food security for the context of the essay and [b] sets out how different examples will be used to form a discussion. There is a clear idea of intent here.

Regional differences in food security exist in two of the world's super powers: China and the USA [a]. In China, there is an ongoing food security issue of providing sufficient food to its vast population of 1.3 billion, spread over a vast area of different climatic types. In regions such as Tibet, where it is dry for 9 months of the year with limited farmland [b] and subsistence agriculture, food is in limited supply. In contrast, the industrial cities to the east, e.g. Beijing and Shanghai, have obesity issues as rising affluence [b] is leading to diets based on dairy, meat and fast foods. A total of 60% of the population of Tibet does not have a healthy diet, with a lack of fresh fruit and vegetables.

In the USA, within individual cities food insecurity exists. In New York, there are stark contrasts between the wealthy residents with access to a wide range of foods and the deprived areas of the Bronx district, where 37% of the population of the district and 1 in 4 children are food insecure due to low income and a lack of economic access to food [b]. Also within the USA, regional differences exist between affluent states such as California and the largest state, Alaska, where indigenous people are suffering food insecurity due to climate effects on their traditional food supply of whale meat. Indigenous groups believe that 'wild resources' are suffering from contamination, e.g. pollution of sea water, contaminating fish [b]. People are being forced into a diet increasingly based on 'bought' food, which is highly priced due to remoteness [b] and of poor nutritional quality, leading to health issues such as obesity — 28% of adults in Alaska are classified as obese. So, food insecurity exists within ACs and EDCs at regional and local scales.

e [a] This is a very good section containing effective and precise factual details about food security issues by contrasting an AC and EDC. [b] It also addresses the issues of physical and economic access to food, as well as quality of food. The paragraph ends with a mini summary of the discussion point.

Poverty often causes food insecurity within many countries. This is particularly the case in LIDCs and EDCs where the income gap is wide. In Kenya, the slum area of Kibera is home to the most disadvantaged groups [a]. Lack of economic access to food is one issue, but also people are unable to take part in urban agriculture schemes as they lack space, although some grow their own food. Senior citizens lack the mobility and money for sufficient food and women put their children first. Different ethnic groups in the USA, such as African-Americans and Latinos have poor food security as this links to low incomes [a].

e [a] This paragraph is more general, referring to differences between broad groups of people and between affluence and poverty. Examples also are quite broad and stated in general rather than specific terms. It does bring out another reason for food security variations within countries but needs analysing in more depth.

The examples discussed show that food security does vary within countries across the development spectrum, such as within urban areas or even between different groups in society. However, the scale of the problem and the causes of the variations in food security vary a great deal. In many ACs, food insecurity is the result of economic access to food. Starvation and famine are not terms used in the context of ACs. However, in these countries there is a growing problem with the quality of food as poor families are often limited in choice and their diet is based on cheap foods which have additives and lots of sugar and salt. In LIDCs such as Sudan there is widespread food insecurity where quantity of food is the problem due to environmental (and human) factors and starvation and famine are common. In Syria, also, there is widespread food insecurity as a result of its civil war [a].

This essay has shown that food insecurity exists within countries to a large extent. It is the causes of the variations that differ and also the scale of the problem differs in terms of numbers affected and severity of the issue [a]. There is a tendency to think that food insecurity is only associated with the most severe cases of lack of physical access, but all countries need to address the issue to a greater or lesser extent [b].

e [a] It is important to finish the essay with a conclusion — often students use a prefix 'in conclusion' or 'this essay has shown that'. [b] The final sentence is interesting as it moves the argument forward and poses a final thought from the student.

e **23/33 marks awarded = equivalent to a B grade.** This essay has scored 7/9 marks (Level 4 in AO1) for knowledge and understanding and 16/24 marks (Level 3 in AO2) for application of knowledge and understanding to interpret, analyse and evaluate geographical issues. With more factual detail to support parts of the middle section and perhaps more about the physical access to sufficient food being a considerable problem within many LIDCs to give balance, the response would go into Level 4 in AO2.

Hazardous Earth

Question 15*

To what extent is the impact of a volcanic eruption related to the type of eruption?   [33 marks]

Student answer

Volcanic eruptions occur when magma from the upper mantle travels upwards to come out at the surface as lava. This can be on land or under the sea along mid-ocean ridges. There are several different types of volcanic eruptions, which depend on several factors.

Magma varies in its chemistry. If it is acidic, it does not flow easily but is sticky. It blocks up the volcano vent, which means that the pressure can build up inside the volcano. When this pressure reaches a critical level, a very explosive eruption, such as a Plinian eruption, can occur, e.g. at Mount St Helens and Krakatoa **a**. When this type of eruption occurs the impacts can be devastating locally but also affect a much wider area and even the whole world.

In the case of Krakatoa (1883) a series of very large explosions blew the cone apart and triggered tsunamis that travelled around the islands of Java and Sumatra. About 36,000 people were killed, most by the tsunami but some by the pyroclastic flows. Tsunami impacts from a volcano can only come if the volcano is at sea level or under the water, as it requires vast amounts of water to be displaced **b**.

---

e The opening paragraph sets the scene simply and suggests that the answer has an appropriate focus. **a** The second paragraph confirms that the student is focused on the question as the factor of magma chemistry is introduced and directly related to eruption type. **b** The impact of an explosive eruption in terms of a tsunami, described in the context of Krakatoa's eruption, and the particular circumstances for a tsunami to arise are helpful comments.

Explosive eruptions can also generate vast amounts of ash as the lava and parts of the volcano are smashed into fine particles by the force of the eruption. The ash can fall quite close to the volcano but can also be carried further away by winds. If the ash is erupted into the upper atmosphere, it can be carried right around the world. The impacts of ash can be environmental, as heavy falls will smother vegetation causing plants to break under the weight, but also blocking sunlight and preventing photosynthesis from happening **a**. Economic impacts can be the destruction of crops and the loss of livestock, the collapse of buildings due to the weight of ash on their roofs and transport disruption **a**. Mount Ontake in Japan erupted in 2014 and killed 63 people. There were also pyroclastic flows and volcanic bombs erupted as well as ash. Air space in the region was closed, as the ash posed a hazard to aircraft **a**. However, ash can also be erupted from effusive eruptions **b**. In Iceland in 2010, a volcano erupted under the ice sheet. The ash that was produced because of the way magma interacted with the melted water was ejected between 6 km and 9 km into the atmosphere. Strong winds then carried it towards Europe, where its impact was the cancellation of about 100,000 flights. This even affected farmers in places such as Kenya, as they couldn't fly their flowers and vegetable to Europe **b**.

e This next paragraph helps move the answer on by strengthening the link between type of eruption and impacts. **a** Comments about ash are well made and set in the context of a real-world examples. **b** It is encouraging to have some evaluation at this point, with the comments about the effusive Icelandic eruption. Given the difficulty of spelling the volcano's name (Eyjafjallajökull), its absence is not an issue, but it is good that other factual details about this eruption are included.

> Effusive eruptions come from basic magma, often through fissures. These can produce vast amounts of basaltic lava, which is relatively free flowing **a**. The Hawaiian hotspot volcanoes are shield volcanoes made of lava, which flows easily. There are some impacts when lava flows over a road and blocks it **b**.
>
> In Iceland, the eruptions tend to be most hazardous when they occur under the ice and can lead to vast amounts of ice melting **a**. When this finds a way out, it floods the landscape. However, as Iceland is sparsely populated, this doesn't often threaten people.

e **a** The comments about effusive eruptions are a useful counterpoint to the previous ones about explosive eruptions. The accurate observations about type of eruption are linked to possible impacts, so keeping the focus on the question, and **b** including some evaluation regarding the influence of population density on impacts.

> As well as the chemistry of the magma influencing the type of eruption and therefore the impacts, other factors are important, such as the area's level of development. This can greatly influence how volcanic hazards can be managed. There are three types of strategy for managing a hazard, which are modify the event, modify the vulnerability and modify the losses **a**.
>
> There is not much that can be done to modify a volcanic eruption. In some ACs there have been attempts to alter the route a lava flow is taking. In Italy, they have tried digging diversion channels and building earth barriers to alter where lava is flowing on Etna **b**.
>
> Nearly all AC volcanoes that are active are monitored, e.g. the USGS monitors the Cascades. This means that if a volcano is about to erupt warnings can be given and people evacuated **b**.
>
> Modifying losses is mainly done in ACs as they have the resources to carry out search and rescue, and have insurance. However, EDCs such as Indonesia monitor some of their volcanoes and can give warnings about possible eruptions **b**.

e These paragraphs offer some convincing analysis of the role of level of development on the impacts of eruptions.

a The knowledge of the ways in which impacts of eruptions might be modified is very good and **b** supported effectively by examples.

It is clear that the chemistry of magma influences the type of eruption, but there are other factors that influence impacts of eruptions. The level of development is important, as this influences the level of resources a country has to manage the volcano hazard.

ⓔ This conclusion is suitable as it clearly refers back to the question and is evaluative.

ⓔ **26/33 marks awarded = A/B boundary.** This essay has scored 8/9 marks (Level 4 in AO1) for knowledge and understanding and 18/24 marks (Level 3 in AO2) for application of knowledge and understanding to interpret, analyse and evaluate geographical issues. With a couple more evaluative comments, such as about level of social and/or political organisation in an area, or the fact that improving knowledge and understanding of volcanic eruptions allows increased modification of vulnerability and or losses, the response would climb into Level 4 in AO2.

Knowledge check answers

1 Natural forcing mechanisms cause climate change. External ones are from outside the Earth, e.g. amount of solar output. Internal forcings operate with the Earth, e.g. amount of carbon dioxide in the atmosphere.

2 Sea-level rise is caused by the thermal expansion of water as the oceans warm. The melting of ice sheets and glaciers causes water to flow from where it is stored as ice on land to the oceans.

3 Even if a gas is not particularly common in the atmosphere, if it has a high global warming potential (GWP), it can seriously impact global warming. Carbon dioxide is not as effective as other gases in trapping energy but because there is so much carbon dioxide, much warming takes place.

4 Ocean acidification will result in population crashes among organisms in the bottom trophic level, thereby affecting all marine ecosystems. Many human populations rely on marine ecosystems for food and other resources, and these will be damaged if these ecosystems are put under severe stress.

5 Mitigation refers to measures aimed at reducing the level of climate change, namely restricting the rise in global temperatures by reducing GHG emissions. Adaptation refers to measures aimed at defending people and environments from the effects of climate change, such as managed realignment of the coastline and the use of GM crops.

6 Disease diffusion is the process by which a disease spreads outwards from its geographical source.

7 Zoonotic diseases are those that can pass between animals and humans, such as rabies.

8 Epidemiological transition is the shift in disease patterns from epidemics of infection to degenerative and man-made diseases as the main cause of morbidity and death. This transition is closely associated with the development of a country over long periods of time.

9 Communicable diseases are infectious diseases that spread from host to host, such as malaria or influenza. Non-communicable diseases cannot spread between people because they are not infectious and not contagious, such as type-2 diabetes or leukaemia.

10 Eradication is complete and permanent worldwide reduction to zero of new cases of a disease through deliberate efforts — no further control measures are required. Elimination is reduction to zero (or a very low defined target rate) of new cases in a defined geographical area — continued measures to prevent re-establishment of disease transmission are required.

11 'Top-down' strategies are the measures used by national governments or international organisations in the control of disease. 'Bottom-up' strategies are attempts to control disease from the 'grass roots', involving and empowering local communities.

12 The halocline and thermocline represent boundary zones in the oceans. Water below these zones is generally unable to rise through them, as it is colder and denser than surface water. Where water from the deep does rise it brings vast quantities of nutrients, which is important for biodiversity.

13 The availability of dissolved nutrients is very important. The deep oceans and regions far from land receive very limited supply of nutrients. Areas closer to land, especially where major rivers flow into the sea, have much higher levels of nutrients, and so NPP can be high.

14 Because there are advantages and disadvantages to the ways in which ocean resources can be used, they are contested among different groups. For example, whale hunting generates strong feelings on both sides of the argument as to whether it should be allowed or not.

15 Some radioactive waste is highly toxic to living organisms, causing genetic mutations, disease and high levels of mortality. It is also remarkably persistent in the environment. Some substances remain dangerously radioactive for thousands of years.

16 As ice cover diminishes, the ability to sail surface vessels around the Arctic Ocean increases. Commercial traffic can exploit the shorter sea routes. Countries bordering the Arctic Ocean are realising the vast potential wealth that exists in the energy and mineral resources. They are keen to claim rights over territory and this brings their respective militaries into closer contact.

17 Greater interconnectedness has increased transnational flows of people, goods and information. New routes and methods of transport have improved access to global food sources. Trade in food and diets has changed as a result. World population growth has also increased the demand for food, and now food can be transported long distances to satisfy this demand.

18 Dietary patterns have changed and as countries become more affluent, consumption shifts from cereals to more expensive food sources, such as dairy and meat and more processed foods. This can lead to more calories consumed than can be worked off through a more sedentary lifestyle, with higher car ownership and more sedentary service sector jobs. Fast-food outlets have also expanded globally, which is another source of food with high calorific content.

19 Climate affects soil characteristics as it affects the rate of development and depth — colder conditions lead to slow development and shallower soils. Climate also affects the movement of minerals through the soil — in very wet climates nutrients are quickly moved through the soil, a process known as leaching. Climate can also lead to salinisation and waterlogging in soils, which leads to soil and land degradation.

20 Neo-Malthusian: the theories of Malthus have gained recent support by demographers known as neo-Malthusians based on recent evidence of famines, wars and water security.

21 Pinch points are disruptions in flow, which can occur in every stage of the food supply chain. They can be political, economic, environmental or technological in nature and can impact at the local, regional and global level.

22 Poverty > increased need to grow more food > more pressure on the land, overgrazing, expansion of cropped areas > reduced vegetation cover > increased soil erosion > desertification.

23 Low precipitation and high evaporation > salts in the soil are brought to the surface > plants intake water but leave salts behind > a salt layer is left, which is toxic to plants and can inhibit water absorption and directly affect plant growth > land is consequently unusable for agriculture.

24 Runoff of nutrients leads to eutrophication, which is excessive algae growth leading to deoxygenation of water and the death of some species, particularly fish.

25 Malnutrition is a shortage of vitamins and essential minerals caused by an unhealthy diet. This can be improved by food retailers providing healthy food at an accessible price and education programmes to help people understand what makes a healthy diet. There can also be overnutrition, where too many calories are consumed — again, the content of processed foods and nutrition education need to be the focus.

26 Global events such as economic recession, food-supply shocks, civil unrest, food riots and concern over the long-term food supply because of global warming have meant that food supply is increasingly affecting and being influenced by political decisions and events.

27 Criticisms of the WTO's role in food security:
- It is mostly controlled by Western nations, especially the USA.
- The rights and democracy of poorer nations are not well represented.
- Some large corporations are successfully reversing environmental protections, attacking them as barriers to trade.
- It isn't doing enough to achieve equitable distribution of food.
- Doha trade talks eventually collapsed due to disagreement between the USA, China and India, who would not reduce tariffs.

28 Costs include: economic leakage of profits back to the parent country; low-cost production can negatively impact small-scale farmers and price them out of the market; sometimes TNCs have more control over food production than national governments; environmental impacts can often be greater; jobs are at risk if the company closes branch plants.

Benefits include: job creation; new ideas and technology brought into the country; the economy benefits from employment through the multiplier effect.

29 Because the asthenosphere is semi-molten, it moves under the influence of convection currents. As it moves, the rigid lithosphere and crust above are dragged across the Earth's surface. This is the basis of plate tectonics.

30 The more acidic lava is, the less free flowing it is. Viscous lava tends to clog the vent, resulting in great pressure building up within the volcano. When the pressure reaches a critical level a very explosive eruption takes place.

31 Liquefaction is where ground (including soil) with a high water content loses its mechanical strength due to being violently shaken during an earthquake. The ground then behaves like a liquid and cannot support structures, which topple and collapse.

32 As population has increased globally, regionally and locally, more and more people are living in seismically active zones. When a tectonic event occurs, more people are affected.

33 Aseismic design is architecture and engineering of structures, including buildings, bridges and tunnels, so that they can withstand a certain degree of earthquake energy.

Index

Note: **Bold** page numbers indicate key terms.